HILTON and WU—A Course in Modern Algebra
HOCHSTADT—Integral Equations
KOBAYASHI and NOMIZU—Foundations of Differential Geometry, Vol. I
KOBAYASHI and NOMIZU—Foundations of Differential Geometry, Vol. II
KRANTZ—Function Theory of Several Complex Variables
LALLEMENT—Semigroups and Combinatorial Applications
LAMB—Elements of Soliton Theory
LAY—Convex Sets and Their Applications
McCONNELL and ROBSON—Noncommutative Noetherian Rings
MELZAK—Invitation to Geometry
NAYFEH—Perturbation Methods
NAYFEH and MOOK—Nonlinear Oscillations
PETRICH—Inverse Semigroups
PIER—Amenable Locally Compact Groups
PRENTER—Splines and Variational Methods
RAO—Measure Theory and Integration
RENELT—Elliptic Systems and Quasiconformal Mappings
RICHTMYER and MORTON—Difference Methods for Initial-Value Problems, 2nd Edition
ROCKAFELLAR—Network Flows and Monotropic Optimization
SCHUMAKER—Spline Functions: Basic Theory
SHAPIRO—Introduction to the Theory of Numbers
SIEGEL—Topics in Complex Function Theory

Volume 1—Elliptic Functions and Uniformization Theory
Volume 2—Automorphic Functions and Abelian Integrals
Volume 3—Abelian Functions and Modular Functions of Several Variables

STAKGOLD—Green's Functions and Boundary Value Problems
STOKER—Differential Geometry
STOKER—Nonlinear Vibrations in Mechanical and Electrical Systems
TURÁN—On A New Method of Analysis and Its Applications
WHITHAM—Linear and Nonlinear Waves
ZAUDERER—Partial Differential Equations of Applied Mathematics

ELLIPTIC SYSTEMS AND QUASICONFORMAL MAPPINGS

ELLIPTIC SYSTEMS AND QUASICONFORMAL MAPPINGS

HEINRICH RENELT
Martin Luther University
Halle–Wittenberg, GDR

A Wiley–Interscience Publication

JOHN WILEY AND SONS

Chichester · New York · Brisbane · Toronto · Singapore

Original title: Quasikonforme Abbildungen und elliptische Systeme
erster Ordnung in der Ebene
(TEUBNER-TEXTE zur Mathematik, Band 46)

British Library Cataloguing in Publication Data:

Renelt, Heinrich
Elliptic systems and quasiconformal mappings.—(Pure and applied mathematics, ISSN 0079-8185).
1. Differential equations. Elliptic
I. Title II. Series
515.3′53 QA377
ISBN 0 471 91731 1

Library of Congress Cataloging-in-Publication Data:

Renelt, Heinrich, 1942–
Elliptic systems and quasiconformal mappings.

(Pure and applied mathematics)
Revised translation of: Quasikonforme Abbildungen und elliptische Systeme erster Ordnung in der Ebene.
'A Wiley-Interscience publication.'
Bibliography: p.
1. Differential equations, Elliptic-Numerical solutions. 2. Analytic functions. 3. Quasiconformal mappings. I. Title. II. Series: Pure and applied mathematics (John Wiley & Sons)
QA374.R42613 1988 515.3′53 87-25347

ISBN 0 471 91731 1

Printed in the German Democratic Republic

Preface

The elliptic systems of first order partial differential equations in the plane

$$f_{\bar{z}} = \nu f_z + \mu \overline{f_z} + Af + B\bar{f} + C \tag{1}$$

contain two special cases of particular interest, namely the systems

$$f_{\bar{z}} = Af + B\bar{f} \tag{2}$$

(the solutions of which are the generalized analytic functions in the sense of Vekua (cf. [V2]) and, simultaneously, pseudoanalytic functions of the first kind in the sense of Bers (cf., e.g. [B1])), and the systems

$$f_{\bar{z}} = \nu f_z + \mu \overline{f_z}, \tag{3}$$

which in real notation read

$$v_y = au_x + bu_y, \qquad -v_x = cu_x + du_y, \tag{3'}$$

with $f = u + iv$, $z = x + iy$, cf. section II.1.

The systems (3) are the main subject of this book. Their solutions, which might be called, in keeping with other notations, generalized analytic functions of the second kind, shall be called (ν, μ)-solutions here. According to the Bers–Nirenberg representation theorem they have the structure 'analytic function composed with a quasiconformal mapping'. Quasiconformal mappings therefore play a crucial role in the treatment of such systems.

Apart from a few minor alterations and the additional section VI.8, the present text is a fairly direct translation of the German edition published in 1982 in the Teubner-Text series. It is thought of as being an introduction. This, however, may not mean that it is easily readable throughout or that it does not contain new results.

With pleasure I express my gratitude to Dr. R. Kühnau for many of his suggestions that have flowed in this text. Also, I wish to thank Dr. H. Kirsten from the English Department at the Martin-Luther-Universität Halle–Wittenberg for his advice in producing this translation. Finally, I would like to express my thanks to Teubner-Verlag Leipzig and to John Wiley & Sons publishing house for their kind cooperation in the production of this volume.

Halle/Saale, May 1987

H. Renelt

Contents

Preface v

0. Auxiliary results from analysis 1

0.1 Notations 1
0.2 Some auxiliary results from measure theory 4
0.3 Some theorems on generalized derivatives 7
0.4 The integral operators P, P_0 and T 13

I. Quasiconformal mappings and solutions of Beltrami systems 23

I.1 The analytic definition of quasiconformal mappings, Beltrami systems 23
I.2 Existence and properties of schlicht Beltrami solutions 25
I.3 Compactness and convergence properties of quasiconformal mappings 32

II. Elliptic systems of partial differential equations of first order in the plane 40

II.1 Conditions and definitions 41
II.2 The representation theorem 43
II.3 Geometrical interpretation of (ν, μ)-solutions 46
II.4 Compactness 53
II.5 Differentiability, a-priori-estimates at interior points 54
II.6 Adjoint systems, the Cauchy integral theorem 62

III. Some function theoretic properties of (ν, μ)-solutions 65

III.1 Zeroes, isolated singularities 65
III.2 Argument principle, invariance of domains, maximum principle 67

III.3 Asymptotic expansions at finite points 69
III.4 Asymptotic expansions at the point ∞, compactness 72

IV. Integral transformations and fundamental solutions **80**

IV.1 Integral operators for (ν, μ)-systems 81
IV.2 Some special cases 84
IV.3 Construction of fundamental solutions by means of the kernels ϕ_1, ϕ_2 86
IV.4 Fundamental solutions in more general cases 88

V. Integral formulae for (ν, μ)-solutions **92**

V.1 The generalized Cauchy integral formula 92
V.2 Relations between fundamental solutions 96
V.3 Cauchy integrals, saltus conditions 99
V.4 Generalization of the Schwarz and Poisson formulae 103

VI. Variational methods for schlicht (ν, μ)-solutions **109**

VI.1 Preliminaries 110
VI.2 The variational formulae for $\varkappa$-conformal boundary variation 116
VI.3 The variational formulae for inner variation 118
VI.4 Two examples of functionals 121
VI.5 Some transformations 122
VI.6 Functionals of Grötzsch-Teichmüller type with respect to $\mathfrak{Q}(\nu, \mu)$ and their extremal functions 128
VI.7 The two examples once more 134
VI.8 A special extremal problem 136

Bibliography **141**

Index **144**

Chapter 0

AUXILIARY RESULTS FROM ANALYSIS

At first, we collect some auxiliary results needed in the following chapters.

0.1 NOTATIONS

Let $\mathbb{R}$ be the set of real numbers, and $\mathbb{C}$ be the finite complex plane, that is, the set of all complex numbers $z = x + iy$, $x, y \in \mathbb{R}$, equipped with the topology induced by the Euclidean distance.

Further, let $\overline{\mathbb{C}} = \mathbb{C} \cup \{\infty\}$ be the full complex plane with the usual topology (cf., e.g. [Y], p. 5).

A set $G \subseteqq \overline{\mathbb{C}}$ is called a domain if G is open and connected. For any $A \subseteqq \overline{\mathbb{C}}$, $\bar{A}$ denotes the closure of A in $\overline{\mathbb{C}}$, ∂A the set of all boundary points of A in $\overline{\mathbb{C}}$, and A^0 the set of all interior points of A. $A \Subset B$ means $\bar{A} \subset B^0$.

For any $A, B \subset \overline{\mathbb{C}}$ with $A \neq \{\infty\}$, $B \neq \{\infty\}$ we write

1.1 $$\operatorname{dist}(A, B) = \inf\{|z - z'|: z \in A, z' \in B\},$$

1.2 $$\operatorname{diam} A = \sup\{|z - z'|: z, z' \in A\}.$$

As usual, for any $a, b \in \mathbb{R}$ we write $[a, b] = \{x: a \leqq x \leqq b\}$, $(a, b) = \{x: a < x < b\}$, $[a, b) = \{x: a \leqq x < b\}$, $(-\infty, a] = \{x: x \leqq a\}$. The terms $(a, b]$, $[a, \infty)$, $(-\infty, +\infty)$ are defined analogously.

By $\mathscr{C}(e_1, e_2)$, e_1 and e_2 any sets $\subseteqq \mathbb{C}$, we denote the set of all the functions defined in $e_1 \cup e_2$ with range $\subseteqq \mathbb{C}$, having the following property: for each $z_0 \in e_1$ and each $\varepsilon > 0$ there exists a $\delta(\varepsilon, z_0, f)$ such that

1.3 $$|f(z) - f(z_0)| < \varepsilon \quad \forall z \in e_2 \quad \text{with} \quad |z - z_0| < \delta(\varepsilon, z_0, f).$$

Of course, $\mathscr{C}(e, e) = \mathscr{C}(e) = \mathscr{C}^0(e)$, the latter in their usual meaning (cf., e.g. [Y], p. 26).

A function $f \in \mathscr{C}(e_1, e_2)$ is called uniformly continuous in $e_1' \subseteqq e_1$, if there is a uniform $\delta(\varepsilon, z_0, f) = \delta(\varepsilon, f)$ for every $z_0 \in e_1'$.

A set M of functions from $\mathscr{C}(e_1, e_2)$ is called equicontinuous in $e_1' \subseteqq e_1$, if there is a $\delta(\varepsilon) > 0$ for every $\varepsilon > 0$ which may be taken as a $\delta(\varepsilon, z_0, f)$ for every $z_0 \in e_1'$ and every $f \in M$.

As usual, a function f defined in a domain $G \subseteqq \mathbb{C}$ is called Hölder continuous in an open set $D \subseteqq G$, if there is a constant K (called the Hölder constant) and a constant $\alpha > 0$ (called the Hölder exponent or, for short, exponent) so that

$$|f(z) - f(z')| \leqq K|z - z'|^\alpha \quad \forall z, z' \in D.$$

A set M of functions defined in a domain G is called Hölder continuous in the open set $D \subseteqq G$ if there are two fixed positive constants K and α so that every $f \in M$ is Hölder continuous with the Hölder constant K and the exponent α in D.

Let the function $f(z) = u(x, y) + iv(x, y)$, $u = \operatorname{Re} f$, $v = \operatorname{Im} f$, $z = x + iy$, be partially differentiable with respect to x and y in the domain $G \subseteqq \mathbb{C}$. We primarily use the formal derivatives f_z, $f_{\bar z}$ of f, defined by

1.4 $$f_z = \frac{\partial f}{\partial z} = \frac{1}{2}(f_x - if_y), \qquad f_{\bar z} = \frac{\partial f}{\partial \bar z} = \frac{1}{2}(f_x + if_y).$$

Of course, for example, f_x means $u_x + iv_x$. In this notation, the Cauchy–Riemann system $u_x = v_y$, $u_y = -v_x$ simply reads $f_{\bar z} = 0$. As to the Jacobian of $f(z) = u(x, y) + iv(x, y)$ we have

1.5 $$u_x v_y - u_y v_x = |f_z|^2 - |f_{\bar z}|^2.$$

Let f be defined in $\mathbb{C}$. We put

1.6 $$\operatorname{supp}_0 f = \{z : z \in \mathbb{C}, f(z) \neq 0\}.$$

Then the support of f is defined by

1.7 $$\operatorname{supp} f = \overline{\operatorname{supp}_0 f} \setminus \{\infty\}.$$

The notations $\mathscr{C}^k(G)$, $\mathscr{C}^\infty(G)$, $\mathscr{C}_0^k(G)$, $\mathscr{C}_0^\infty(G)$, where G is any open set $\subseteqq \mathbb{C}$, and k is any non-negative integer, are used here in their usual meaning (cf., e.g. [Y], p. 26). For instance, a function f belongs to $\mathscr{C}_0^\infty(G)$ if we have $\operatorname{supp} f \Subset G$ and if all its derivatives of any order exist and are continuous in G. Instead of $\mathscr{C}^k(\mathbb{C})$, $\mathscr{C}^\infty(\mathbb{C})$, $\mathscr{C}_0^k(\mathbb{C})$, $\mathscr{C}_0^\infty(\mathbb{C})$ we only write $\mathscr{C}^k$, $\mathscr{C}^\infty$, $\mathscr{C}_0^k$, $\mathscr{C}_0^\infty$, respectively.

Unless otherwise stated, measure always means the two-dimensional Lebesgue measure, and to be measurable (integrable) always means to be measurable (integrable) with respect to the two-dimensional Lebesgue measure. By $|A|$ we denote the outer measure of an arbitrary set $A \subseteqq \mathbb{C}$ and simultaneously the measure of a measurable set A.

Let E be a measurable set $\subseteqq \mathbb{C}$. As usual, $L_p(E)$ with $1 \leqq p < \infty$ means the Banach space of all functions f with $|f(z)|^p$ integrable in E, equipped with the norm

1.8 $$\|f\|_{L_p(E)} = \Big(\iint_E |f(z)|^p \, d\sigma\Big)^{1/p} = \Big(\iint_E |f(z)|^p \, d\sigma_z\Big)^{1/p}.$$

(Integrability means existence and finiteness of the integral.)

Further, $L_\infty(E)$ denotes the Banach space of all the functions defined a.e. (almost everywhere) in E, which are measurable and essentially bounded in E, under the norm

1.8′ $$\|f\|_{L_\infty(E)} = \operatorname{ess\,supp}\{|f(z)| : z \in E\} = \inf\{\sup\{|f(z)| : z \in E \setminus A\} : |A| = 0\}.$$

We will say that the function f belongs to $L_{p,\mathrm{loc}}(E)$ if $f \in L_p(D)$ for every measurable set $D \Subset E$. Instead of $L_p(\mathbb{C})$ and $L_{p,\mathrm{loc}}(\mathbb{C})$ we write L_p and $L_{p,\mathrm{loc}}$, respectively.

Measurability of a set $G \subseteq \overline{\mathbb{C}}$ means that $G \setminus \{\infty\}$ is a measurable set, and $L_p(G)$, $L_{p,\mathrm{loc}}(G)$ are to be abbreviations of $L_p(G \setminus \{\infty\})$, $L_{p,\mathrm{loc}}(G \setminus \{\infty\})$, respectively, if $\infty \in G$. Accordingly,

$$\iint_G f \, d\sigma \quad \text{always means} \quad \iint_{G\setminus\{\infty\}} f \, d\sigma,$$

$|G|$ is always defined by $|G \setminus \{\infty\}|$, and a.e. in G really means a.e. in $G \setminus \{\infty\}$.

We will say that an object $\mathscr{A}$ (e.g. a function or a sequence of functions, etc.) possesses a certain property E locally in $D \subseteq \overline{\mathbb{C}}$, if $\mathscr{A}$ has this property in each measurable set $D' \Subset D$. In case $D = \mathbb{C}$ we will only say that $\mathscr{A}$ locally has the property E. For instance, a family M of functions f is called locally equicontinuous, if M is equicontinuous in each bounded measurable subset of $\mathbb{C}$.

Let $\varphi(z)$ be any function defined and $\neq 0$ in $\{0 < |z| < r\}$, r a positive number. Then $o(\varphi(z))$, $O(\varphi(z))$ denote the usual Bachmann–Landau symbols, i.e. we have

1.9 $$\limsup_{z\to 0} \left|\frac{O(\varphi(z))}{\varphi(z)}\right| < \infty, \qquad \lim_{z\to 0} \frac{o(\varphi(z))}{\varphi(z)} = 0,$$

unless otherwise stated (namely, e.g., $o(\varphi(z))/\varphi(z) \to 0$ in the sense of convergence in L_p if $\varphi(z) \to 0$ in L_p).

Now let φ be any function defined and $\neq 0$ in $\{|z| > r\}$, r any positive number. In this connection $o(\varphi(z))$ and $O(\varphi(z))$ are to represent functions for which

1.10 $$\limsup_{z\to \infty} \left|\frac{O(\varphi(z))}{\varphi(z)}\right| < \infty, \qquad \lim_{z\to \infty} \frac{o(\varphi(z))}{\varphi(z)} = 0.$$

A function f defined in a domain G is called differentiable at a point $z_0 \in G$ if there are constants a and b such that

1.11 $$f(z) = a(z - z_0) + b(\overline{z - z_0}) + o(z - z_0).$$

Of course, then $f_z(z_0)$, $f_{\bar z}(z_0)$ exist, and $f_z(z_0) = a$, $f_{\bar z}(z_0) = b$.

A set M of functions shall be called equidifferentiable at $z_0 \in G$ if for every $\varepsilon > 0$ there is a $\delta > 0$ such that

$$\left|\frac{o(z - z_0)}{z - z_0}\right| < \varepsilon \quad \text{if } 0 < |z - z_0| < \delta, \text{ for every } f \in M,$$

with $o(z - z_0) = o_f(z - z_0)$ belonging to the f at the time according to 1.11.

If, in an equation, the left-hand (right-hand) side is a definition of the term on the other side, we occasionally use the symbol '$=:$' ('$:=$') in place of '$=$'.

0.2 SOME AUXILIARY RESULTS FROM MEASURE THEORY

Let us now collect some measure theoretic results.

2.1 Egoroff's theorem. *Let $f_1, f_2, f_3, \ldots$ and g be functions measurable and finite a.e. in the measurable set A with $|A| < \infty$. If $\lim_{n\to\infty} f_n(z) = g(z)$ at almost all $z \in A$, then for each $\varepsilon > 0$ there exists a closed set $F \subseteqq A$ with $|A \setminus F| < \varepsilon$ where the convergence of the f_n to g is uniform.*

For a proof cf., e.g. [F1], p. 77 (and note that for each measurable set, B, of finite measure, and each $\varepsilon > 0$ there is a closed set $F \subseteqq B$ with $|B \setminus F| < \varepsilon$).

Let τ be any regular Borel measure (abbreviated RBM) over $\mathbb{C}$ (in the terminology used in [F1], p. 61). That means, τ is a function with range in $\{t: 0 \leqq t \leqq +\infty\}$, defined on the class $2^{\mathbb{C}}$ of the subsets of $\mathbb{C}$, with the following additional properties:

(I) $\tau(A) \leqq \sum_{i=1}^{\infty} \tau(B_i)$ if $B_i \subseteqq \mathbb{C}$, $A \subseteqq \bigcup_{i=1}^{\infty} B_i$,

(II) every open subset of $\mathbb{C}$ is τ-measurable, and $\tau(\emptyset) = 0$,

(III) for every $A \subseteqq \mathbb{C}$ there exists a Borel set $B \supseteqq A$ such that $\tau(B) = \tau(A)$.

As usual, any $A \subseteqq \mathbb{C}$ is called τ-measurable if

2.2 $$\tau(T) = \tau(T \setminus A) + \tau(T \cap A) \quad \forall T \subseteqq \mathbb{C}.$$

Properties (I) and (II) imply τ-measurability of every Borel set $B \subseteqq \mathbb{C}$ (cf., e.g. [F1], 2.2.1, 2.1.3).

We call τ a finite RBM if τ is RBM and additionally

(IV) $\tau(A) < \infty$ for every bounded $A \subset \mathbb{C}$.

According to the terminology in [F1], a finite RBM is called absolutely continuous if

2.3 $$\tau(E) = 0 \text{ for every } E \subset \mathbb{C} \text{ with } |E| = 0$$

(this property is occasionally also called 'locally absolutely continuous').

We have:

2.4 *A finite RBM τ is absolutely continuous if and only if for each $\varepsilon > 0$ and each bounded open set $W \subset \mathbb{C}$ there exists a $\delta(\varepsilon, W)$ such that $\tau(e) < \varepsilon$ for every $e \subset W$ with $|e| < \delta(\varepsilon, W)$.*

Proof. The one direction is obvious. To prove the other we consider a sequence $A_n \subset \mathbb{C}$ with $\tau(A_n) \geqq \varepsilon > 0$, $|A_n| < 2^{-n}$, $A_n \subset W$. Since the Lebesgue measure $|\cdot|$ is a finite RBM, there exist Borel sets B_n with $|B_n| = |A_n|$, $W \supseteqq B_n \supseteqq A_n$; consequently, $\tau(B_n) \geqq \tau(A_n) \geqq \varepsilon$. Let $C_k = \bigcup_{n=k}^{\infty} B_n$, $B = \bigcap_{k=1}^{\infty} C_k$. Then we have

$$|B| \leqq |C_k| \leqq \sum_{n=k}^{\infty} |B_n| \leqq 2^{-k+1} \text{ for every } k = 1, 2, \ldots, \text{ hence } |B| = 0.$$

On the other hand, the τ-measurability of the C_k implies

$$\tau(B) = \lim_{k\to\infty} \tau(C_k) \geqq \limsup_{k\to\infty} \tau(B_k) \geqq \varepsilon$$

(note that $\tau(C_1) < \infty$ and cf. [F1], 2.1.3(5)). Assertion 2.4 is proved.

A family $\mathscr{T}$ consisting of finite RBM τ is called absolutely equicontinuous if there exists, for each $\varepsilon > 0$ and each open and bounded set $W \subset \mathbb{C}$, a $\delta(\varepsilon, W) > 0$ such that

2.5 $\quad \tau(e) < \varepsilon$ for every $\tau \in \mathscr{T}$ and every $e \subset W$ with $|e| < \delta(\varepsilon, W)$.

Let $Q(r)$ be any closed square having sides of length r parallel to the coordinate axes. Let V denote the set of all such squares, $0 < r < \infty$. We fix any $z \in \mathbb{C}$ and consider a sequence $Q(r_n) \in V$ with $z \in Q(r_n)$, $n = 1, 2, \ldots$, and $r_n \to 0$ as $n \to \infty$. If the limit

2.6
$$\lim_{n\to\infty} \frac{\tau(Q(r_n))}{|Q(r_n)|} = \lim_{n\to\infty} \tau(Q(r_n))\, r_n^{-2}$$

exists for every such sequence $Q(r_n)$, it is called the derivative $\tau'(z)$ of τ at z.

We shall later need the following result.

2.7 Lebesgue's theorem. *Let τ be a finite RBM. Then $\tau'(z)$ exists a.e. in $\mathbb{C}$, $\tau'(z)$ is a measurable function, and for every measurable set A it holds*

(I) $\tau(A) \geqq \iint_A \tau'(z)\, d\sigma_z$.

Equality holds for every measurable set $A \subseteqq \mathbb{C}$ if and only if τ is absolutely continuous, and then every measurable set is also τ-measurable.

For the proof we refer to [F1], 2.9.1, 2.9.7, 2.9.16, 2.9.2.

Let $w(z)$ be a homeomorphism of a domain $G \subseteqq \mathbb{C}$ onto a domain $G_w \subseteqq \mathbb{C}$. By

2.8 $\quad \tau(e) := |w(e \cap G')|$, where G' is any open set $\Subset G$,

we obtain a finite RBM. Indeed, $B \subseteqq G$ is a Borel set if and only if its image $w(B) \subseteqq G_w$ is a Borel set (cf. [F1], 2.2.14), and $|\cdot|$ is a finite RBM.

The homeomorphism $w(z)$ is called absolutely continuous, if every τ corresponding to $w(z)$ according to 2.8 for any open set $G' \Subset G$ is absolutely continuous (in the sense of 2.3). A set of homeomorphisms is called absolutely equicontinuous, if for each open set $G' \Subset G$ the corresponding set of finite RBM is absolutely equicontinuous.

2.9 *If $w(z)$ is an absolutely continuous homeomorphism of the domain G, then $w(A)$ is measurable for every measurable $A \subset G$.*

Indeed, let $w(z)$ be an absolutely continuous homeomorphism of the domain G, and A be any measurable set, $A \Subset G$. There exists a Borel set $B \supseteqq A$ with $|B| = |A|$, and, of course, we may additionally assume that $B \subseteqq G' \Subset G$ with G' an open set. By the measurability of A we have $|B \setminus A| = 0$. Because of the absolute continuity of $w(z)$ it holds $0 = |w(B \setminus A)| = |w(B) \setminus w(A)|$, i.e. $w(B) \setminus w(A)$ is a null set. This implies measurability of $w(A)$ since $w(B)$ is a Borel set. Because any measurable $A \subset G$ can be represented by the union of a sequence of measurable $A_n \Subset G$, assertion 2.9 follows.

Further we have

2.10 Theorem. ([L/V], p. 130). *Let the orientation-preserving homeomorphism $w(z)$ of the domain G be differentiable at $z_0 \in G$. For any τ that corresponds to $w(z)$ according to* 2.8 *with a G' containing z_0, the derivative $\tau'(z_0)$ exists and is equal to $|w_z(z_0)|^2 - |w_{\bar{z}}(z_0)|^2$.*

As to the proof we refer to [L/V], p. 131.

Concerning differentiability of homeomorphisms, Gehring and Lehto established the following result.

2.11 Theorem of Gehring and Lehto. *Let $w(z)$ be a homeomorphism of the domain $G \subseteqq \mathbb{C}$ onto the domain $G_w \subseteqq \mathbb{C}$, having finite partial derivatives w_z, $w_{\bar{z}}$ a.e. in G. Then $w(z)$ is differentiable a.e. in G.*

As to the proof, we refer to [L/V], pp. 128–130.

Furthermore, we need the following result on the transformation of variables in integrals, proved in [L/V], pp. 121 and 131–132.

2.12 Transformation rule. *Let $w(z)$ be an orientation-preserving homeomorphism of the domain G onto the domain $G_w = w(G)$ with the following properties:*

(1) *$w(z)$ and its inverse mapping $z(w)$ are absolutely continuous,*
(2) *$w(z)$ is differentiable a.e. in G.*

Then for any $f \in L_1(G_w)$ and any measurable $A \subseteqq G$ it holds

(I) $$\iint_{w(A)} f(w)\, d\sigma_w = \iint_A f \circ w(z)\, (|w_z|^2 - |w_{\bar{z}}|^2)\, d\sigma_z.$$

Note, in view of later applications that, by 2.12(1), theorem 2.7, 2.12(2) and theorem 2.10, the Jacobian of $w(z)$ is positive a.e. in G, hence $z(w)$ is differentiable a.e. in G_w. This means that under the same conditions as in 2.12

2.12' $$\iint_A f(z)\, d\sigma_z = \iint_{w(A)} f \circ z(w)\, (|z_w|^2 - |z_{\bar{w}}|^2)\, d\sigma_w$$

for any $f \in L_1(G)$ and any measurable $A \subseteqq G$.

0.3 SOME THEOREMS ON GENERALIZED DERIVATIVES

Let f and g be two functions from $L_{1,\mathrm{loc}}(G)$, G an open set $\subseteqq \mathbb{C}$, so that

3.1 $$\iint f\varphi_z \, d\sigma = -\iint g\varphi \, d\sigma \quad \forall \varphi \in \mathscr{C}_0^\infty(G)$$

(an integral with no integration domain indicated is to be taken over the whole support of its integrand). Then g is called the generalized derivative f_z of f in G. Correspondingly, a function h from $L_{1,\mathrm{loc}}(G)$ is said to be the generalized derivative $f_{\bar{z}}$ of f in G if

3.2 $$\iint f\varphi_{\bar{z}} \, d\sigma = -\iint h\varphi \, d\sigma \quad \forall \varphi \in \mathscr{C}_0^\infty(G).$$

Because $\varphi_z + \varphi_{\bar{z}} = \varphi_x$ we have $\iint f\varphi_x \, d\sigma = -\iint (h + g)\, \varphi \, d\sigma$. This, and the corresponding relation for φ_y, yields

3.3 $$f_z + f_{\bar{z}} = f_x, \qquad \mathrm{i}(f_z - f_{\bar{z}}) = f_y,$$

where f_x, f_y are the generalized derivatives of f defined in a way analogous with 3.1, 3.2. The relations 3.3 are, of course, in keeping with 1.4.

These generalized derivatives are, as elements of $L_{1,\mathrm{loc}}(G)$, uniquely determined, that is, two functions g_1, g_2 satisfying, say, the relation 3.1 with the same f may differ only on a null set (because of the density of $\mathscr{C}_0^\infty(G')$ in $L_1(G')$).

We put

$$W_p^1(G) = \{f\colon f \text{ has generalized derivatives } f_z, f_{\bar{z}} \text{ in } G, \text{ and } f, f_z, f_{\bar{z}} \in L_p(G)\},$$
$$1 \leqq p < \infty.$$

Since $L_p(G)$ is a Banach space and because convergence with respect to the norm in $L_p(G)$ implies weak convergence in $L_p(G)$, $W_p^1(G)$ endowed with the norm

3.4 $$\|f\|_{W_p^1(G)} = \|f\|_{L_p(G)} + \|f_z\|_{L_p(G)} + \|f_{\bar{z}}\|_{L_p(G)}$$

is a Banach space. Moreover, we have:

3.5 *Let f_n be a sequence of functions from $W_p^1(G) \cap \mathscr{C}(G)$, $1 < p < \infty$, f_n converging to f locally uniformly in G as $n \to \infty$, and $\|f_n\|_{W_p^1(G)} \leqq K =$ constant $\forall n = 1, 2, \ldots$ Then $f \in W_p^1(G) \cap \mathscr{C}(G)$, $\|f\|_{W_p^1(G)} \leqq K$, and $f_{nz}, f_{n\bar{z}}$ converge weakly in $L_p(G)$ to $f_z, f_{\bar{z}}$, respectively.*

Indeed, the limit function f is, of course, continuous in G and belongs to $L_p(G)$, because

$$K \geqq K_1 := \limsup_{n\to\infty} \|f_n\|_{L_p(G)} \geqq \lim_{n\to\infty} \|f_n\|_{L_p(G')} = \|f\|_{L_p(G')}$$

for any measurable $G' \Subset G$. Further, for any $\varphi \in \mathscr{C}_0^\infty(G)$ we have

$$\iint f_{nz}\varphi \, d\sigma = -\iint f_n\varphi_z \, d\sigma \to -\iint f\varphi_z \, d\sigma, \qquad \iint f_{n\bar{z}}\varphi \, d\sigma \to -\iint f\varphi_{\bar{z}} \, d\sigma$$

as $n \to \infty$. By the weak compactness of bounded sets in $L_p(G)$ there are $h, g \in L_p(G)$ for a subsequence $f_{n'}$ of f_n to hold $f_{n'z} \rightharpoonup g$, $f_{n'\bar{z}} \rightharpoonup h$ as $n' \to \infty$ ('$\rightharpoonup$' denotes weak convergence). Consequently,

$$\iint g\varphi \, d\sigma = -\iint f\varphi_z \, d\sigma \quad \text{and} \quad \iint h\varphi \, d\sigma = -\iint f\varphi_{\bar{z}} \, d\sigma \quad \forall \varphi \in \mathscr{C}_0^\infty(G).$$

This means $g = f_z$, $h = f_{\bar z}$. Further, from

$$\iint f_{nz}\varphi \, d\sigma \to \iint g\varphi \, d\sigma, \qquad \iint f_{n\bar z}\varphi \, d\sigma \to \iint h\varphi \, d\sigma \quad \forall \varphi \in \mathscr{C}_0^\infty(G)$$

under observing that $\mathscr{C}_0^\infty(G)$ is dense in $L_q(G)$, $p^{-1} + q^{-1} = 1$, and that $\|f_{nz}\|_{L_p(G)}$, $\|f_{n\bar z}\|_{L_p(G)}$ are bounded, we conclude that $f_{nz} \rightharpoonup f_z$, $f_{n\bar z} \rightharpoonup f_{\bar z}$ even for the whole sequence f_n. Moreover, we can choose a subsequence $f_{n'}$ such that

$$\lim_{n'\to\infty} \|f_{n'}\|_{L_p(G)} = K_1, \qquad \lim_{n'\to\infty} \|f_{n'z}\|_{L_p(G)} =: K_2, \qquad \lim_{n'\to\infty} \|f_{n'\bar z}\|_{L_p(G)} =: K_3$$

and $K_1 + K_2 + K_3 \leqq K$. Since the norms of the weak limits $f_z, f_{\bar z}$ satisfy the same bounds as the $f_{n'z}, f_{n'\bar z}$ do (cf., e.g. [Y], p. 120) we obtain $\|f\|_{W_p^1(G)} \leqq K_1 + K_2 + K_3 \leqq K$, too. Assertion 3.5 is proved. Further, let

$$W^1_{p,\mathrm{loc}}(G) = \{f : f \in W_p^1(G') \text{ for every domain } G' \Subset G\}.$$

In place of $W_p^1(\mathbb{C})$, $W^1_{p,\mathrm{loc}}(\mathbb{C})$ we again only write W_p^1, $W^1_{p,\mathrm{loc}}$, respectively.

In connection with generalized derivatives the so-called mollifiers play a role. Let $m_1(z)$ be a function defined in $\mathbb{C}$ with the following properties:

3.6 (I) $m_1(z) \in \mathscr{C}^\infty$, (II) $m_1(z) \geqq 0 \ \forall z \in \mathbb{C}$, (III) $m_1(z) = m_1(|z|) \ \forall z \in \mathbb{C}$, (IV) $m_1(z) = 0$ for $|z| \geqq 1$, (V) $\iint m_1(z)\, d\sigma = 1$.

For instance, we can take $m_1(z) = e\pi^{-1}(|z|^2 - 1)^{-2}\, e^{1/(|z|^2-1)}$ for $|z| < 1$, $m_1(z) = 0$ otherwise in $\mathbb{C}$.

For any positive number ϱ we put

3.7 $$m_\varrho(z) = m_1(z/\varrho)\,\varrho^{-2}.$$

Then for any $f \in L_p(G)$, $1 \leqq p < \infty$, we define

3.8 $$f_\varrho(z) = \iint f(t)\, m_\varrho(t - z)\, d\sigma_t = \iint f(t + z)\, m_\varrho(t)\, d\sigma_t,$$

while putting $f = 0$ in $\mathbb{C} \setminus G$. Concerning f_ϱ we have the following result.

3.9 Theorem. *Let $m_1(z)$ satisfy the conditions* 3.6 (I)–(V), *and let $f \in L_p(G)$, $1 \leqq p < \infty$. Then it holds*

(I) *$f_\varrho \in \mathscr{C}^\infty$ for every $\varrho > 0$,*

(II) *$f_\varrho \in L_p(G)$, $\|f_\varrho\|_{L_p(G)} \leqq \|f\|_{L_p(G)}$ and $\lim_{\varrho\to 0} \|f_\varrho - f\|_{L_p(G)} = 0$.*

(III) *If, additionally, f is continuous in $D \subseteqq G$, then $f_\varrho(z) \to f(z)$ uniformly in every $D' \Subset D$ as $\varrho \to 0$.*

(IV) *If, additionally, f has a generalized derivative $f_{\bar z} \in L_p(G)$ in G, then $(f_\varrho)_{\bar z} = (f_{\bar z})_\varrho$ in $G_\varrho := \{z : z \in G, \operatorname{dist}(\{z\}, \partial G) > \varrho\}$, and a corresponding result holds with f having a generalized derivative f_z.*

As to the proof we refer to, e.g. [S5], pp. 180–182, 290.

Functions with generalized derivatives have an important property, called absolute continuity in the sense of Tonelli. Originally, this property (which will be described in 3.10) was even the starting point for their definition.

3.10 Theorem. *Let $w(z) \in W^1_{1,\mathrm{loc}}(G) \cap \mathscr{C}(G)$, $R := \{z = x + iy : a \leqq x \leqq b, c \leqq y \leqq d\} \subset G$, $R_x(y) := \{z = x + iy : a \leqq x \leqq b\}$, $R_y(x) := \{z = x + iy : c \leqq y \leqq d\}$. For almost every $y \in [c, d]$ the restriction $w(z)|_{R_x(y)}$ is an absolutely* *continuous function of x. Likewise, the restriction $w(z)|_{R_y(x)}$ is an absolutely continuous function of y for almost every $x \in [a, b]$. Moreover, the (classical) partial derivatives w_x, w_y exist a.e. in G and coincide with the corresponding generalized derivatives a.e. in G.*

Proof. By $\iint\limits_R (|w| + |w_x|) \, d\sigma < \infty$ in conjunction with Fubini's theorem we obtain

$$\int_a^b (|w| + |w_x|) \, dx < \infty \quad \text{for almost every } y \in [c, d],$$

where w_x denotes the generalized derivative of w with respect to x in G. For the approximation w_ϱ of w according to 3.8, we have, by 3.9,

$$\iint\limits_R (|w_\varrho - w| + |(w_\varrho)_x - w_x|) \, d\sigma \to 0 \quad \text{as } \varrho \to 0.$$

Once again by Fubini's theorem, we have

$$g_\varrho(y) := \int_a^b (|w_\varrho - w| + |(w_\varrho)_x - w_x|) \, dx < \infty \quad \text{a.e. in } [c, d],$$

$g_\varrho(y) \in L_1([c, d])$, and $\int_c^d g_\varrho(y) \, dy \to 0$ if $\varrho \to 0$.

Since convergence in the mean implies pointwise convergence a.e. of a subsequence, there exists a sequence ϱ_n tending to zero as $n \to \infty$ such that $g_{\varrho_n}(y) \to 0$ pointwise a.e. in $[c, d]$ as $n \to \infty$. The latter implies uniformly absolute continuity (and, of course, uniform boundedness) in $[a, b]$ of $w_{\varrho_n}(z)|_{R_x(y)}$ for every $y \in [c, d] \setminus N$, N a linear null set. Indeed, for any set of disjoint intervals $(x_j, x_j') \subset [a, b]$, $j = 1, 2, \ldots$, we have

$$\sum_j |w_{\varrho_n}(x_j' + iy) - w_{\varrho_n}(x_j + iy)| \leqq \sum_j \int_{x_j}^{x_j'} |w_{\varrho_n x}(x + iy)| \, dx$$
$$\leqq \sum_j \int_{x_j}^{x_j'} |w_x| \, dx + \int_a^b |w_{\varrho_n x} - w_x| \, dx.$$

From this and from the uniform convergence of the $w_{\varrho_n}(z)$ to $w(z)$, especially in $R_x(y)$ (cf. theorem 3.9(III) above), we obtain absolute continuity of $w(z)|_{R_x(y)}$ in $[a, b]$ for every $y \in [c, d] \setminus N$. Consequently, the (classical) derivative $w_{x,cl}$ of w exists at almost all points $x \in [a, b]$ for each $y \in [c, d] \setminus N$. We still have to prove that $w_{x,cl} = w_x$ a.e. in R. By changing over to a further subsequence we may additionally assume that $\lim\limits_{n \to \infty} w_{\varrho_n x}(z) = w_x(z)$ a.e. in $[a, b]$ for every $y \in [c, d] \setminus N^*$, where N^* is a null set, and without loss o generality $N = N^*$. Furthermore,

$$\lim_{n \to \infty} \int_a^b w_{\varrho_n} \varphi_x \, dx = \int_a^b w \varphi_x \, dx = -\int_a^b w_{x,cl} \varphi \, dx \quad \forall y \in [c, d] \setminus N, \quad \varphi \in \mathscr{C}_0^\infty(R),$$

and

$$\lim_{n \to \infty} - \int_a^b w_{\varrho_n} \varphi_x \, dx = \lim_{n \to \infty} \int_a^b w_{\varrho_n x} \varphi \, dx = \int_a^b w_x \varphi \, dx + \lim_{n \to \infty} \int_a^b (w_{\varrho_n x} - w_x) \, \varphi \, dx$$
$$= \int_a^b w_x \varphi \, dx,$$

hence

$$\int_a^b w_{x,\mathrm{cl}}\varphi \,\mathrm{d}x = \int_a^b w_x\varphi \,\mathrm{d}x \quad \forall\varphi \in \mathscr{C}_0^\infty(R) \quad\text{and}\quad \forall y \in [c, d] \setminus N.$$

It is not difficult to see that the latter implies

$$w_{x,\mathrm{cl}}(z) = w_x(z) \quad \text{a.e. in } [a, b] \text{ for every } y \in [c, d] \setminus N$$

(note that any bounded measurable function $f(x)$, say, with compact support in $[a, b]$, can be approximated by a sequence f_n of $\mathscr{C}_0^\infty$-functions such that $\lim_{n\to\infty} f_n(x) = f(x)$ a.e. in $[a, b]$, all f_n are uniformly bounded, and $f_n(x)$ can be complemented by a suitable $\mathscr{C}_0^\infty$-factor $g(y)$ such that $f_n(x)\,g(y)$ belongs to $\mathscr{C}_0^\infty(R)$). Thus we have

$$\int_c^d \left(\int_a^b |w_{x,\mathrm{cl}}| \,\mathrm{d}x\right) \mathrm{d}y = \int_c^d \left(\int_a^b |w_x| \,\mathrm{d}x\right) \mathrm{d}y = \iint_R |w_x| \,\mathrm{d}\sigma_z,$$

and, by Fubini's theorem, the existence of the iterated integral on the left-hand side implies integrability of $w_{x,\mathrm{cl}}$ over R. Hence $w_{x,\mathrm{cl}}\varphi + w\varphi_x \in L_1(R)$ for every $\varphi \in \mathscr{C}_0^\infty(R)$. Thus

$$\iint_R (w_{x,\mathrm{cl}}\varphi + w\varphi_x) \,\mathrm{d}\sigma = \int_c^d \left(\int_a^b (w_{x,\mathrm{cl}}\varphi + w\varphi_x) \,\mathrm{d}x\right) \mathrm{d}y = \int_c^d \left(\int_a^b (w\varphi)_{x,\mathrm{cl}} \,\mathrm{d}x\right) \mathrm{d}y = 0,$$

and this means that $w_{x,\mathrm{cl}}$ is also a generalized derivative of w in R in the sense of 3.1 to 3.3. From the uniqueness of generalized derivatives we finally obtain $w_x = w_{x,\mathrm{cl}}$ a.e. in R. The corresponding considerations are also valid for w_y. Since $R \Subset G$ is arbitrary, theorem 3.10 is proved.

As an obvious consequence of theorem 3.10 we have the following.

3.11 Corollary. *Let $w(z)$ be differentiable a.e. in the domain G,*

$$w(z + h) - w(z) = a(z)\, h + b(z)\, \bar{h} + o(h) \quad \textit{for almost all } z \in G,$$

and let $w(z) \in W_{1,\mathrm{loc}}^1(G) \cap \mathscr{C}(G)$. Then $w_z(z) = a(z)$, $w_{\bar{z}}(z) = b(z)$ a.e. in G.

The theorem of Gehring and Lehto reads, in conjunction with theorem 3.10, as follows.

3.12 Theorem. *Let $w(z)$ be a homeomorphism of the domain G, and $w(z) \in W_{1,\mathrm{loc}}^1(G)$. Then $w(z)$ is differentiable a.e. in G.*

We now come to Weyl's lemma in its complex form due to Ahlfors [A1].

3.13 Weyl's lemma. *Let $w(z) \in L_{1,\mathrm{loc}}(G)$, $w(z)$ have the generalized derivative $w_{\bar{z}} \equiv 0$ in G, G a domain. Then, after possible alteration on a null set, $w(z)$ is an analytic function in G.*

Proof. We fix any $\varrho_0 > 0$. Because of 3.9(IV) and $w_{\bar{z}} \equiv 0$, $w_\varrho(z)$ is analytic in each component of G_{ϱ_0} for every ϱ with $0 < \varrho < \varrho_0$. Moreover we have

3.14 $$w_{\varrho_1}(z) = w_{\varrho_2}(z) \quad \text{in } G_{\varrho_0} \quad \forall \varrho_1, \varrho_2 \in (0, \varrho_0).$$

Indeed, let $\varrho > 0$ satisfy $\varrho_1 + \varrho < \varrho_0$, $\varrho_2 + \varrho < \varrho_0$. The functions w_{ϱ_1}, w_{ϱ_2} are analytic in $G_{\varrho_0 - \varrho}$, and from 3.6(III), 3.6(V), and the mean value property of analytic functions we infer

3.15 $$w_{\varrho_j}(z) = \iint w_{\varrho_j}(t + z)\, m_\varrho(t)\, d\sigma_t \quad \forall z \in G_{\varrho_0}, \quad j = 1, 2.$$

The latter is

$$= \iint \left(\iint w(s+t)\, m_{\varrho_j}(s - z)\, d\sigma_s\right) m_\varrho(t)\, d\sigma_t = \iint \left(\iint w(s+t)\, m_\varrho(t)\, d\sigma_t\right) m_{\varrho_j}(s - z)\, d\sigma_s$$
$$= \iint w_\varrho(s)\, m_{\varrho_j}(s - z)\, d\sigma_s = w_\varrho(z) \quad \text{in } G_{\varrho_0},$$

for the reasons as in 3.15, and this gives 3.14. Furthermore, by 3.9(II) there are sequences $\varrho_n \to 0$ such that $w_{\varrho_n} \to w$ a.e. in G as $n \to \infty$. Together with 3.14 this yields $w_\varrho(z) = w(z)$ a.e. in G_{ϱ_0}. The lemma is proved.

We also need the following

3.16 Lemma ([A2], chapter V B). *Let $P(z), Q(z) \in W^1_{1,\mathrm{loc}}(G) \cap \mathscr{C}(G)$, where G is a simply connected domain, and let*

$$P_{\bar z} = Q_z \quad \text{a.e. in } G.$$

Then there is an $F \in \mathscr{C}^1(G)$ such that $F_z = P$, $F_{\bar z} = Q$.

Proof. By hypothesis and 3.9(IV) we have

$$0 = \oint_{\mathfrak{C}} P_\varrho\, dz + Q_\varrho\, d\bar z \quad \text{for any rectifiable closed path } \mathfrak{C} \subset G,$$

if $\varrho > 0$ is sufficiently small. In conjunction with 3.9(III) this implies $\oint_{\mathfrak{C}} P\, dz + Q\, d\bar z = 0$. Consequently, $\int_{z_0}^{z} P\, dz + Q\, d\bar z$ defines a function F with the desired properties.

Let us also recall

3.17 Green's formula. *Let $f \in W^1_{p,\mathrm{loc}}(G)$, $h \in W^1_{q,\mathrm{loc}}(G)$, $f, h \in \mathscr{C}(G)$, $p^{-1} + q^{-1} = 1$, and D be an n-fold connected domain whose boundary ∂D consists of n rectifiable Jordan curves $\mathfrak{C}_1, \mathfrak{C}_2, \ldots, \mathfrak{C}_n$, where $\mathfrak{C}_1$ is positively oriented and contains $\mathfrak{C}_2, \ldots, \mathfrak{C}_n$ in its interior, $\mathfrak{C}_2, \ldots, \mathfrak{C}_n$ are to be negatively oriented, and $D \Subset G$. If h is of bounded variation on ∂D, then we have*

(I) $$\int_{\partial D} f\, dh = \iint_D (f_x h_y - f_y h_x)\, d\sigma.$$

Proof. The validity of Green's formula for $\mathscr{C}^\infty$-functions may be considered to be well-known, i.e. we have

3.18 $$\int_{\partial D} f_\varrho\, dh_{\varrho'} = \iint_D [(f_\varrho)_x (h_{\varrho'})_y - (f_\varrho)_y (h_{\varrho'})_x]\, d\sigma.$$

In view of 3.9(III) we obtain

$$\lim_{\varrho' \to 0} \int_{\partial D} f_\varrho\, dh_{\varrho'} = -\lim_{\varrho' \to 0} \int_{\partial D} h_{\varrho'}\, df_\varrho = -\int_{\partial D} h\, df_\varrho = \int_{\partial D} f_\varrho\, dh,$$

and because of 3.9 (II) (convergence with respect to the norm implies weak convergence) and 3.9 (IV), the right-hand side of 3.18 tends to

$$\iint_D [(f_\varrho)_x h_y - (f_\varrho)_y h_x] \, d\sigma$$

as $\varrho' \to 0$. For the same reasons it follows

$$\int_{\partial D} f \, dh = \lim_{\varrho \to 0} \int_{\partial D} f_\varrho \, dh = \lim_{\varrho \to 0} \iint_D [(f_x)_\varrho \, h_y - (f_y)_\varrho \, h_x] \, d\sigma = \iint_D (f_x h_y - f_y h_x) \, d\sigma,$$

which had to be proved.

In complex notation, 3.17 (I) reads

3.19
$$\int_{\partial D} f \, dh = 2i \iint_D (f_{\bar z} h_z - f_z h_{\bar z}) \, d\sigma.$$

By means of Green's formula in conjunction with 2.7 and 3.10 (for details cf. [L/V], p. 150) one obtains the following useful theorem.

3.20 Theorem. *Let $w(z)$ be a homeomorphism of the domain $G \subseteqq \mathbb{C}$, $w(z) \in W^1_{2,\mathrm{loc}}(G)$. Then $w(z)$ is absolutely continuous.*

We now collect some rules about generalized derivatives. By means of the Hölder inequality

3.21 $\|fg\|_{L_t(G)} \leqq \|f\|_{L_r(G)} \cdot \|g\|_{L_s(G)} \quad \forall f \in L_r(G), \quad \forall g \in L_s(G),$

where $r^{-1} + s^{-1} = t^{-1} \leqq 1$, and approximations according to theorem 3.9, we obtain the following.

3.22 Product rule. *Let $f \in W^1_r(G)$, $g \in W^1_s(G)$, $r^{-1} + s^{-1} = t^{-1} = 1$. Then $fg \in W^1_t(G)$ and*

(I) $(fg)_z = f_z g + f g_z, \qquad (fg)_{\bar z} = f_{\bar z} g + f g_{\bar z}.$

If $f, g \in W^1_{r,\mathrm{loc}}(G) \cap \mathscr{C}(G)$ or $f \in W^1_{r,\mathrm{loc}}(G)$, $g \in \mathscr{C}^1(G)$, then $fg \in W^1_{r,\mathrm{loc}}(G)$ and (I) *holds.*

In the same way we obtain

3.23 Chain rule I. *Let $1 \leqq p < \infty$, and either $w(z) \in W^1_{p,\mathrm{loc}}(G) \cap \mathscr{C}(G)$, $f(\cdot) \in \mathscr{C}^1(G_w)$ and $w(G) \subseteqq G_w$, or $w(z) \in W^1_{p,\mathrm{loc}}(G)$ and $f(\cdot) \in \mathscr{C}^1_0(\mathbb{C})$. Then $f \circ w(z) \in W^1_{p,\mathrm{loc}}(G)$ and*

(I) $(f \circ w)_z = f_w w_z + f_{\bar w} \bar w_z, \qquad (f \circ w)_{\bar z} = f_w w_{\bar z} + f_{\bar w} \overline{w_z}.$

A little more involved is

3.24 Chain rule II. *Let $f(\cdot) \in W^1_q(G_w) \cap L_s(G_w)$, $1 < q < 2$, and $w(z)$ be an orientation-preserving homeomorphism of the domain $G \subseteqq \mathbb{C}$ onto $G_w \subseteqq \mathbb{C}$ with the properties*

(I) *$w(z)$ is absolutely continuous, $w(z) \in W^1_1(G)$,*

(II) *the inverse mapping $z(w)$ of $w(z)$ belongs to $W^1_p(G_w)$, $p^{-1} + q^{-1} = 1$, and $(p/2)^{-1} + s^{-1} = 1$.*

Then $f \circ w(z) \in W_1^1(G)$, *and* 3.23 (I) *holds.*

Proof. To start with let us note that, by 3.12, 3.20 and $p > 2$, the homeomorphism $w(z)$ satisfies all the conditions needed in the transformation rules for integrals 2.12 and 2.12′. Now, by chain rule I we have

3.25 $$\iint [(f_\varrho)_w w_z + (f_\varrho)_{\bar{w}} \bar{w}_z] \varphi(z)\, d\sigma_z = -\iint f_\varrho \circ w(z) \varphi_z(z)\, d\sigma_z$$

for every $\varphi \in \mathscr{C}_0^\infty(G)$. The latter is, by 2.12′,

$$= -\iint f_\varrho(w) \varphi_z(z(w)) (|z_w|^2 - |z_{\bar{w}}|^2)\, d\sigma_w,$$

and, by 3.9 (II), this tends to

3.26 $$-\iint f(w) \varphi_z(z(w)) (|z_w|^2 - |z_{\bar{w}}|^2)\, d\sigma_w = -\iint f \circ w(z) \varphi_z(z)\, d\sigma_z$$

as $\varrho \to 0$. The Jacobian of $w(z)$ is positive a.e. in G (cf. the remarks in note 2.12). Hence, by 3.11, 3.12, and Cramer's rule we have

3.27 $$z_w = \frac{\overline{w_z}}{|w_z|^2 - |w_{\bar{z}}|^2}, \qquad z_{\bar{w}} = \frac{-w_{\bar{z}}}{|w_z|^2 - |w_{\bar{z}}|^2},$$

$$w_z = \frac{\overline{z_w}}{|z_w|^2 - |z_{\bar{w}}|^2}, \qquad w_{\bar{z}} = \frac{-z_{\bar{w}}}{|z_w|^2 - |z_{\bar{w}}|^2}$$

a.e. in G, G_w, respectively. Consequently, the left-hand side in 3.25 is, once more by 2.12′,

$$= \iint [(f_\varrho)_w \overline{z_{\bar{w}}} - (f_\varrho)_{\bar{w}} \bar{z}_w] \varphi \circ z(w)\, d\sigma_w,$$

and by theorem 3.9 this tends to

3.28 $$\iint [f_w \overline{z_{\bar{w}}} - f_{\bar{w}} \bar{z}_w] \varphi \circ z(w)\, d\sigma_w = \iint (f_w w_z + f_{\bar{w}} \bar{w}_z) \varphi(z)\, d\sigma_z$$

as $\varrho \to 0$. Therefore, the right-hand sides in 3.26 and 3.28 are equal to one another for every $\varphi \in \mathscr{C}_0^\infty(G)$, and this means validity of the first relation in 3.23 (I). In the same way the second relation follows. Chain rule II is proved.

The same considerations as in the proof of chain rule II, in conjunction with theorem 3.20, lead to the following variant of chain rule II.

3.29 *Let* $f(\cdot) \in W^1_{q,\mathrm{loc}}(G_w)$, $1 \leqq q < \infty$, *and let* $w(z)$ *be an orientation-preserving homeomorphism of the domain* G *onto* G_w, $w(z) \in \mathscr{C}^1(G)$, $z(w) \in \mathscr{C}^1(G_w)$. *Then* $f \circ w(z) \in W^1_{q,\mathrm{loc}}(G)$, *and* 3.23 (I) *holds.*

0.4 THE INTEGRAL OPERATORS P, P_0 AND T

We write $L^{p,q}(G) := L_p(G) \cap L_q(G)$, $L^{p,q} := L^{p,q}(\mathbb{C})$, where $p, q \in [1, \infty)$, and G is an open set $\subseteqq \mathbb{C}$.

$L^{p,q}(G)$ equipped with the norm

4.1 $$\|f\|_{L^{p,q}(G)} = \max\{\|f\|_{L_p(G)}, \|f\|_{L_q(G)}\}$$

is a Banach space. Furthermore, let

4.2 $$Ph(z) = -\frac{1}{\pi}\iint \frac{h(t)}{t - z}\, d\sigma_t \quad \text{for } h \in L^{p,q},\quad 1 \leqq q < 2 < p < \infty,$$

and

4.2′ $$P_0 g(z) = -\frac{1}{\pi}\iint g(t)\left[\frac{1}{t-z} - \frac{1}{t}\right] d\sigma_t \quad \text{for } g \in L_p,\quad 2 < p < \infty.$$

Obviously, both integrals in 4.2, 4.2′ are well-defined for every $z \in \mathbb{C}$.

4.3 Theorem. *Let* $2 < p < \infty$, $1 < q < 2$, $h \in L^{p,q}$, $g \in L_p$. *Then*

(I) $|Ph(z)| \leqq K_{p,q}\|h\|_{L^{p,q}} \quad \forall z \in \mathbb{C}$,

(II) $|P_0 g(z)| \leqq K_p \|g\|_{L_p} \cdot |z|^{1-(2/p)} \quad \forall z \in \mathbb{C}$,

(III) $|Ph(z_1) - Ph(z_2)| \leqq K_p \|h\|_{L_p} |z_1 - z_2|^{1-(2/p)}$,

$|P_0 g(z_1) - P_0 g(z_2)| \leqq K_p \|g\|_{L_p} |z_1 - z_2|^{1-(2/p)} \quad \forall z_1, z_2 \in \mathbb{C}$,

(IV) $\lim_{z\to\infty} Ph(z) = 0$.

Here we may set

$$K_{p,q} = \inf\left\{\left\|\frac{1}{t}\right\|_{L_{p'}(\{|t|<R\})} + \left\|\frac{1}{t}\right\|_{L_{q'}(\{|t|>R\})} : 0 < R < \infty\right\},$$

$$K_p = \pi^{-1} \cdot \|1/(t(t-1))\|_{L_{p'}}, \qquad p^{-1} + p'^{-1} = 1, \qquad q^{-1} + q'^{-1} = 1.$$

Proof. The assertion (I) follows by splitting the range of integration $\mathbb{C}$ into $\{|t - z| < R\}$ and $\{|t - z| > R\}$ and applying the Hölder inequality. Furthermore, we have

4.4 $$|P_0 g(z)| \leqq \frac{1}{\pi}\iint |g| \cdot |z| \cdot |(t-z)\,t|^{-1}\, d\sigma_t$$
$$\leqq \frac{|z|}{\pi}\|g\|_{L_p}\left\|\frac{1}{(t-z)\,t}\right\|_{L_{p'}} = K_p\|g\|_{L_p} \cdot |z|^{1-(2/p)},$$

first if $z \neq 0$, and because $P_0 g(0) = 0$ also for $z = 0$. Assertion (II) is proved.

Assertion (III) is a consequence of (II). Indeed, we have

$$|P_0 g(z_1) - P_0 g(z_2)| = \left|\frac{1}{\pi}\iint g(t)\left[\frac{1}{t-z_1} - \frac{1}{t-z_2}\right] d\sigma_t\right|$$
$$= \left|\frac{1}{\pi}\iint g(t+z_2)\left[\frac{1}{t-(z_1-z_2)} - \frac{1}{t}\right] d\sigma_t\right|,$$

and this is, because of (II),

$$\leqq K_p\|g(t+z_2)\|_{L_p} |z_1 - z_2|^{1-(2/p)} = K_p\|g\|_{L_p} |z_1 - z_2|^{1-(2/p)}.$$

Moreover, if $h \in L^{p,q}$, then also $P_0 h$ is defined, and we have

4.5 $$P_0 h(z) = Ph(z) - Ph(0).$$

Consequently,

$$|Ph(z_1) - Ph(z_2)| = |P_0 h(z_1) - P_0 h(z_2)| \leqq K_p\|h\|_{L_p} |z_1 - z_2|^{1-(2/p)}.$$

Assertion (III) is proved.

To prove (IV) let $|z| =: 2r > 2$, $K(z, s) := \{t : |t - z| < s\}$, $q' = q/(q-1)$, $p' = p/(p-1)$. Then $1 < p' < 2$, $q' > 2$, and

$$\left|\iint \frac{h(t)}{t-z}\,d\sigma_t\right| \leqq \left|\iint_{K(0,3r)\setminus K(z,r)}\right| + \left|\iint_{K(z,r)\setminus K(z,1)}\right| + \left|\iint_{K(z,1)}\right| + \left|\iint_{|t|>3r}\right|$$

$$\leqq \|h\|_{L_q}(8\pi r^2)^{1/q'}\cdot r^{-1} + \|h\|_{L_q(\{|t|>r\})}\cdot \|1/t\|_{L_{q'}(\{|t|>1\})}$$

$$+ \|h\|_{L_p(\{|t|>r\})}\cdot \|1/t\|_{L_{p'}(\{|t|<1\})} + \|h\|_{L_q(\{|t|>3r\})}\,\|1/t\|_{L_{q'}(\{|t|>r\})}.$$

This tends to zero as $r \to \infty$. Theorem 4.3 is proved.

Furthermore, for any $g \in \mathscr{C}_0^1$, $Pg(z)$ has (classical) derivatives with respect to z and $\bar{z}$:

4.6 $$(Pg(z))_{\bar{z}} = -\frac{1}{\pi}\iint \frac{g_{\bar{t}}(t)}{t-z}\,d\sigma_t = g(z)$$

and

4.7 $$(Pg(z))_z = -\frac{1}{\pi}\iint \frac{g_t(t)}{t-z}\,d\sigma_t = \lim_{\varepsilon\to+0} -\frac{1}{\pi}\iint_{|t-z|>\varepsilon} \frac{g(t)}{(t-z)^2}\,d\sigma_t.$$

The first equations in 4.6 and 4.7 follow by considering the difference quotients and applying the mean value theorem and Lebesgue's theorem on bounded convergence. To prove the second equation in 4.7 we apply 3.19 with $h = \bar{z}$ and obtain

4.8 $$-\frac{1}{\pi}\iint \frac{g_t(t)}{t-z}\,d\sigma_t$$

$$= \lim_{\varepsilon\to+0}\left[-\frac{1}{\pi}\iint_{|t-z|>\varepsilon}\left\{\frac{g_t(t)}{t-z} - \frac{g(t)}{(t-z)^2}\right\}d\sigma_t - \frac{1}{\pi}\iint_{|t-z|>\varepsilon}\frac{g(t)}{(t-z)^2}\,d\sigma_t\right]$$

$$= \lim_{\varepsilon\to+0}\left[-\frac{1}{2\pi i}\oint_{|t-z|=\varepsilon}\frac{g(t)}{t-z}\,d\bar{t} - \frac{1}{\pi}\iint_{|t-z|>\varepsilon}\frac{g(t)}{(t-z)^2}\,d\sigma_t\right] = Tg(z)$$

where

4.9 $$Tg(z) = \lim_{\varepsilon\to+0} -\frac{1}{\pi}\iint_{|t-z|>\varepsilon}\frac{g(t)}{(t-z)^2}\,d\sigma_t,$$

because

$$\lim_{\varepsilon\to 0}\oint_{|t-z|=\varepsilon}\frac{g(t)}{t-z}\,d\bar{t} = 0.$$

The second equation in 4.6 follows in an analogous way. The assertions 4.6 and 4.7 are proved.

The operator T, defined so far only in $\mathscr{C}_0^1$, is called the two-dimensional Hilbert transformation.

Since $P_0 g$ and Pg differ only by a constant if $g \in L^{p,q} \supset \mathscr{C}_0^1$, it holds

4.10 $$(P_0 g)_z = (Pg)_z, \quad (P_0 g)_{\bar{z}} = (Pg)_{\bar{z}} \quad \forall g \in \mathscr{C}_0^1.$$

From 4.3(III) and 4.7 we infer:

4.11 *For every $g \in \mathscr{C}_0^1$ $Tg(z)$ is Hölder continuous in $\mathbb{C}$ with any Hölder exponent* <1.

Furthermore, 4.6 and 4.7 imply

4.12 $(Tg)_{\bar z} = (P(g_z))_{\bar z} = g_z, \qquad (Tg)_z = (P(g_z))_z = T(g_z) = P(g_{zz}) \quad \forall g \in \mathscr{C}_0^2.$

Let $(h, g) := \iint h\bar{g}\,\mathrm{d}\sigma$ be the scalar product in L_2. We have

4.13 $$(Th, Tg) = (h, g) \quad \forall h, g \in \mathscr{C}_0^2,$$

because

$$\begin{aligned}\iint Th \cdot \overline{Tg}\,\mathrm{d}\sigma &= \iint (Ph)_z\,(\overline{Pg})_{\bar z}\,\mathrm{d}\sigma = -\iint Ph \cdot (\overline{Pg})_{\bar z z}\,\mathrm{d}\sigma = -\iint Ph \cdot (\overline{P(g_z)})_z\,\mathrm{d}\sigma \\ &= -\iint Ph \cdot \overline{(P(g_z))_{\bar z}}\,\mathrm{d}\sigma = -\iint Ph \cdot \overline{g_z}\,\mathrm{d}\sigma = -\iint Ph \cdot \bar{g}_{\bar z}\,\mathrm{d}\sigma \\ &= \iint \bar{g} \cdot (Ph)_{\bar z}\,\mathrm{d}\sigma = \iint h\bar{g}\,\mathrm{d}\sigma.\end{aligned}$$

As $\mathscr{C}_0^2$ is dense in L_2 and because T, by 4.13, is linear and bounded in $\mathscr{C}_0^2$ with respect to the L_2-norm, T can be extended to the whole of L_2 with

4.14 $$(Th, Tg) = (h, g), \quad \text{especially} \quad \|Th\|_{L_2} = \|h\|_{L_2}, \quad \forall g, h \in L_2.$$

As usual, we shall write this operator as an integral,

4.15 $$Tg(z) = -\frac{1}{\pi}\iint \frac{g(t)}{(t-z)^2}\,\mathrm{d}\sigma_t.$$

This notation has certain advantages because it describes the nature of T in an adequate manner, cf. theorem 4.16 below.

By general results of Calderón and Zygmund (cf. [S6] and corresponding literature cited there) the following theorem holds.

4.16 Theorem. *Let $\varepsilon > 0$, $h \in L_p$, $1 < p < \infty$. Then*

$$T_\varepsilon h(z) := -\frac{1}{\pi}\iint_{|t-z|>\varepsilon} \frac{h(t)}{(t-z)^2}\,\mathrm{d}\sigma_t$$

is defined $\forall z \in \mathbb{C}$, belongs to L_p again, and

(I) $\|T_\varepsilon h\|_{L_p} \leqq A_p\|h\|_{L_p}$, *where the constant A_p does not depend on ε,*

(II) *for each $h \in L_p$ there exists exactly one $g \in L_p$, which we call Th, such that $\|Th - T_\varepsilon h\| \to 0$ as $\varepsilon \to 0$ and, consequently, $\|Th\|_{L_p} \leqq A_p\|h\|_{L_p}$,*

(III) $\lim_{\varepsilon\to 0} T_\varepsilon h(z) = Th(z)$ *pointwise a.e. in $\mathbb{C}$.*

For the proof we refer to [S6], chap. II, § 4.

Let $C(p)$ denote the norm of this operator in L_p. By a general theorem of M. Riesz on convexity (cf., e.g. [S/W], chap. V.1, or [A2], chap. V D) we have

4.17 Theorem. *The* log $C(p)$ *is a convex function of $\dfrac{1}{p}$ for $\dfrac{1}{p} \in (0, 1)$. Consequently, $C(p)$ is a continuous function of p in $(1, \infty)$.*

Because of 4.14

4.18 $$C(2) = 1.$$

Furthermore, we need the following theorem.

4.19 Theorem. *Let* $h \in L^{p,q}$, $g \in L_p$, $1 < q < 2 < p < \infty$. *Then* Ph *and* $P_0 g$ *have generalized derivatives with respect to* $z, \bar{z}$, *and*

(I) $(Ph)_z = Th$, $(Ph)_{\bar{z}} = h$,

(II) $(P_0 g)_z = Tg$, $(P_0 g)_{\bar{z}} = g$.

Proof. For every $h \in L^{p,q}$ there exists a sequence $h_n \in \mathscr{C}_0^\infty$ with $h_n \to h$ in $L^{p,q}$ as $n \to \infty$. By 4.7, 4.11 we have

4.20 $$\iint Ph_n \cdot \varphi_z \, d\sigma_z = -\iint Th_n \cdot \varphi \, d\sigma \quad \forall \varphi \in \mathscr{C}_0^\infty.$$

The left-hand side tends, by theorem 4.3(I), to $\iint Ph \cdot \varphi_z \, d\sigma_z$ as $n \to \infty$, and the right-hand side tends, by theorem 4.16(I), (II), to $-\iint Th \cdot \varphi \, d\sigma$. This proves the first equation in 4.19(I). Also, taking into account theorem 4.3 the remaining assertions follow analogously.

With $p, q \geqq 1$ let

4.21 $L_{p,q}(G) = \{f = f_1 + f_2 : f_1 \in L_p(G), f_2 \in L_q(G)\}$, $\quad L_{p,q} = L_{p,q}(\mathbb{C})$.

The norm

4.22 $\|f\|_{L_{p,q}(G)} = \inf\{\|f_1\|_{L_p(G)} + \|f_2\|_{L_q(G)} : f = f_1 + f_2, f_1 \in L_p(G), f_2 \in L_q(G)\}$

makes $L_{p,q}(G)$ a Banach space (cf., e.g. [T], p. 18).

4.23 Theorem. *Let* $p, q \in (1, \infty)$. (I) *Then every linear continuous functional* $l(f)$ *in* $L_{p,q}(G)$ *has the form*

$$l(f) = \iint_G fg \, d\sigma \quad \text{with } g \in L^{p',q'}(G), \qquad \|l\| = \|g\|_{L^{p',q'}(G)},$$

where $p^{-1} + p'^{-1} = 1$, $q^{-1} + q'^{-1} = 1$, *and*

(II) $L_{p,q}(G)$ *is reflexive.*

(III) $L_2(G) \subset \bigcap_{\substack{1 \leqq q < 2 \\ 2 < p < \infty}} L_{p,q}(G)$, $\|f\|_{L_{p,q}(G)} \leqq \|f\|_{L_2(G)}^{2/p} + \|f\|_{L_2(G)}^{2/q}$ $\quad \forall f \in L_2(G)$.

(IV) $L^{p,q}(G) \subseteqq L^{r,s}(G)$, $L_{r,s}(G) \subseteqq L_{p,q}(G)$, *if* $1 \leqq q \leqq r \leqq s \leqq p < \infty$.

Proof. Because $L_p(G) \subset L_{p,q}(G)$ and $\|f\|_{L_p(G)} \geqq \|f\|_{L_{p,q}(G)}$, the restriction of $l(\cdot)$ to $L_p(G)$ is linear and continuous in $L_p(G)$. This implies the existence of just one $g_p \in L_{p'}(G)$, $p'^{-1} + p^{-1} = 1$, such that

$$l(f) = \iint_G fg_p \, d\sigma \quad \forall f \in L_p(G).$$

In the same manner we have

$$l(f) = \iint_G fg_q \, d\sigma \quad \forall f \in L_q(G)$$

with just one $g_q \in L_{q'}(G)$, $q'^{-1} + q^{-1} = 1$. Consequently,

$$\iint_G fg_p \, d\sigma = \iint_G fg_q \, d\sigma \quad \forall f \in L^{p,q}(G),$$

hence $g_p = g_q =: g$ a.e. in G. This means $g \in L^{p',q'}(G)$ and $l(f) = \iint\limits_G fg \, d\sigma$ $\forall f \in L_{p,q}(G)$, the latter by the linearity of $l(\cdot)$. Further, the Hölder inequality yields

4.24 $\quad |l(f)| \leqq \|g\|_{L^{p',q'}(G)} \cdot \|f\|_{L_{p,q}(G)}, \quad \text{hence} \quad \|l\| \leqq \|g\|_{L^{p',q'}(G)}.$

Without loss of generality we may assume $\|g\|_{L^{p',q'}(G)} = \|g\|_{L_{p'}(G)}$. Since the restriction of l to $L_p(G)$ has the norm $\|g\|_{L_{p'}(G)}$, there is, for each positive ε, an $f \in L_p(G)$ of norm $\neq 0$ and

$$\left| \iint\limits_G gf \, d\sigma \right| \geqq (1 - \varepsilon) \|g\|_{L_{p'}(G)} \cdot \|f\|_{L_p(G)} \geqq (1 - \varepsilon) \|g\|_{L^{p',q'}(G)} \cdot \|f\|_{L_{p,q}(G)}.$$

The arbitrariness of $\varepsilon > 0$ together with 4.24 gives $\|l\| = \|g\|_{L^{p',q'}(G)}$. Assertion 4.23(I) is proved.

By the Eberlein–Shmulyan theorem (cf. [Y], p. 141), $L_{p,q}(G)$ is reflexive if and only if it holds:

4.25 *Every bounded set in $L_{p,q}(G)$ is weakly compact.*

Proof. Let a sequence f_n bounded in $L_{p,q}(G)$ be given, $\|f_n\|_{L_{p,q}(G)} \leqq M$ $\forall n = 1, 2, \ldots$ For each f_n there are $f_{1n} \in L_p(G)$, $f_{2n} \in L_q(G)$ such that $f_n = f_{1n} + f_{2n}$ and $\|f_{1n}\|_{L_p(G)} + \|f_{2n}\|_{L_q(G)} \leqq M + 1$. By changing over to a subsequence we may assume, without loss of generality, that, by the weak compactness of bounded sets in $L_p(G)$ and $L_q(G)$, the f_{1n} converge weakly to $h_1 \in L_p(G)$ and f_{2n} to $h_2 \in L_q(G)$. Because of 4.23(I) $f_n = f_{1n} + f_{2n}$ converges weakly to $h_1 + h_2$ in $L_{p,q}(G)$. This proves 4.25 and 4.23(II).

To prove 4.23(III) we divide $f \in L_2(G)$ into f_1 and f_2 such that

4.26 $\quad f_1 = \begin{cases} f & \forall z \in G \text{ with } |f(z)| < 1, \\ 0 & \text{otherwise in } \mathbb{C} \end{cases} \qquad f_2 = \begin{cases} f & \forall z \in G \text{ with } |f(z)| \geqq 1 \\ 0 & \text{otherwise in } \mathbb{C}. \end{cases}$

Then for any p, q with $1 \leqq q < 2 < p < \infty$ we have

4.27 $\quad \iint |f_1|^p \, d\sigma \leqq \iint |f_1|^2 \, d\sigma \leqq \|f\|^2_{L_2(G)}, \ \iint |f_2|^q \, d\sigma \leqq \iint |f_2|^2 \, d\sigma \leqq \|f\|^2_{L_2(G)},$

hence $f \in L_{p,q}(G)$, and the inequality in 4.23(III) holds. Moreover, $(z - z_0)^{-1} \in L_{p,q}(G)$ for all the p, q considered, but $(z - z_0)^{-1} \notin L_2(G)$ if z_0 is a point of density of G. Assertion 4.23(III) is proved.

The assertions in 4.23(IV) follow analogously with 4.27 by means of splitting f as in 4.26. Theorem 4.23 is proved.

Our next aim is to extend P, P_0 to an operator P^* defined on $L_{p,q}$ for $1 < q < 2 < p < \infty$ such that

4.28 $\quad (P^*h)_z = Th, \qquad (P^*h)_{\bar{z}} = h \quad \forall h \in L_{p,q}, \quad 1 < q < 2 < p < \infty.$

Here Th is defined in a unique way for every $h \in L_{p,q}$ by $Th = Th_1 + Th_2$ where $h = h_1 + h_2$, $h_1 \in L_p$, $h_2 \in L_q$.

Let $\omega_A(z)$ be the characteristic function of the set $A \subseteq \mathbb{C}$, i.e. $\omega_A(z) = 1$ if $z \in A$ and $= 0$ otherwise. For any function h defined in A we put $h(z)\,\omega_A(z) = 0$ if $z \notin A$.

We write

4.29 $$P_G h(z) = P(h\omega_G)(z) = -\frac{1}{\pi}\iint\limits_G \frac{h(t)}{t-z}\,d\sigma_t,$$

$$P_{0G}g(z) = P_0(g\omega_G)(z) = -\frac{1}{\pi}\iint\limits_G g(t)\left[\frac{1}{t-z}-\frac{1}{t}\right]d\sigma_t,$$

$$T_G f(z) = T(f\omega_G)(z) = -\frac{1}{\pi}\iint\limits_G \frac{f(t)}{(t-z)^2}\,d\sigma_t = -\frac{1}{\pi}\iint \frac{f(t)\,\omega_G(t)}{(t-z)^2}\,d\sigma_t,$$

where $h \in L^{p,q}(G)$, $g \in L_p(G)$, $f \in L_{p,q}(G)$, $1 < q < 2 < p < \infty$. Here we have to interpret the last integral in 4.29 as a Cauchy principal value existing, according to theorem 4.16, for almost all $z \in \mathbb{C}$. The penultimate integral in 4.29 only means a different notation for the last.

To start with let G be any bounded domain. We now define the operator P_G in $L_q(G)$ for $1 < q \leqq 2$.

4.30 *Definition.*

$$P_G h(z) := T_G(th(t))(z) - zT_G h(z)$$

$$= -\frac{1}{\pi}\iint\limits_G \frac{th(t)}{(t-z)^2}\,d\sigma_t + \frac{z}{\pi}\iint\limits_G \frac{h(t)}{(t-z)^2}\,d\sigma_t \quad \text{for } h \in L_q(G), \quad 1 < q \leqq 2.$$

Hereby $P_G h$ is defined a.e. in $\mathbb{C}$, and by the boundedness of T, P_G is a bounded operator on $L_q(G)$ into $L_q(G')$ for any bounded measurable G'. Furthermore, we have

4.31 Theorem ([V2], p. 40). *P_G is a bounded operator on $L_q(G)$ into $L_r(G)$ for every bounded domain G, every $q \in (1, 2]$ and every r satisfying $1 \leqq r < 2q/(2-q)$, that is, we have for these G, q, r,*

(I) $\|P_G h\|_{L_r(G)} \leqq M_{q,r,G}\|h\|_{L_q(G)}$,

where $M_{q,r,G}$ depends only on q, r, G. Moreover,

(II) $(P_G h)_{\bar z} = h\omega_G$, $(P_G h)_z = T_G h$ in $\mathbb{C}$ $\quad\forall h \in L_q(G)$, $q \in (1, 2]$.

Proof. Let $f \in \mathscr{C}_0^\infty(G)$. Then $P_G f(z)$ is an ordinary integral for every $z \in \mathbb{C}$, which is in keeping with the above definition. By the Hölder inequality 3.21 we have, initially for $q < r$,

4.32 $$|P_G f(z)| \leqq \frac{1}{\pi}\iint \frac{|f(t)|}{|t-z|}\,d\sigma_t = \frac{1}{\pi}\iint \frac{|f(t)|^{q/r}}{|t-z|^\beta}\,\frac{|f(t)|^{1-(q/r)}}{|t-z|^{1-\beta}}\,d\sigma_t$$

$$\leqq \frac{1}{\pi}\,\|\,|f(t)|^{q/r}\,|t-z|^{-\beta}\,\|_{L_r(G)}\,\|\,|f(t)|^{1-(q/r)}\,|t-z|^{-1+\beta}\|_{L_s(G)},$$

where

4.33 $r^{-1} + s^{-1} = 1, \quad r\beta < 2, \; s(1 - \beta) < 2, \quad$ hence $\; s = r/(r - 1),$
$$(2/r) - 1 < \beta < 2/r,$$

where β is any number such that 4.33 is satisfied. Further,

$$\| |f(t)|^{1-(q/r)} \, |t - z|^{-(1-\beta)} \|_{L_s(G)} \leqq \| |f(t)|^{1-(q/r)} \|_{L_k(G)} \, \| |t - z|^{-(1-\beta)} \|_{L_m(G)},$$

$k^{-1} + m^{-1} = s^{-1}$, and because we want to obtain an estimate in terms of the L_q-norm of f we put $k(1 - (q/r)) = q$, i.e.

4.34 $k = rq/(r - q), \quad m = q/(q - 1), \quad (1 - \beta)\, m < 2,$ thus $(2/q) - 1 < \beta$.

Let $M(\lambda, G) := \sup \left\{ \iint_G |t - z|^{-\lambda} \, d\sigma_t : z \in G \right\}$. Then we have

4.35 $$|P_G f(z)|^r \leqq \frac{1}{\pi^r} \left(\iint \frac{|f(t)|^q}{|t - z|^{\beta r}} \, d\sigma_t \right) \|f\|_{L_q(G)}^{r-q} \, [M(\lambda, G)]^{r(q-1)/q},$$

when putting $\lambda = (1 - \beta)\, q/(q - 1) < 2$. Integration of both sides (with respect to z) and Fubini's theorem then give

4.36 $$\|P_G f\|_{L_r(G)}^r \leqq \|f\|_{L_q(G)}^{r-q} \, [M(\lambda, G)]^{r(1-(1/q))} \, M(\lambda', G) \, \|f\|_{L_q(G)}^q \, \pi^{-r}$$
$$= \pi^{-r} M(\lambda', G) \, [M(\lambda, G)]^{r(1-(1/q))} \, \|f\|_{L_q(G)}^r, \quad \lambda' = \beta r < 2,$$

and this holds for any $r > q$ as long as there exists a β satisfying the last inequalities both in 4.33 and 4.34. This is true if $(2/q) - 1 < 2/r$, that is, if $r < 2q/(2 - q)$. Thus, 4.31 (I) is valid for any $h \in \mathscr{C}_0^\infty(G)$ and any $r \in (q, 2q/(2 - q))$. By the Hölder inequality and the boundedness of G, 4.31 (I) then also holds for $1 \leqq r \leqq q$. Further, since $\mathscr{C}_0^\infty(G)$ is dense in $L_q(G)$ and T_G is continuous on $L_q(G)$, for any $h \in L_q(G)$ there is a sequence $h_n \in \mathscr{C}_0^\infty(G)$, so that $h_n \to h$ and $P_G h_n \to P_G h$, each with respect to the L_q-norm. The pointwise convergence of a subsequence of the $P_G h_n$ to $P_G h$ and Fatou's lemma (cf. [F1], p. 84), in conjunction with the validity of 4.31 (I) for every h_n, then give the validity of 4.31 (I) also for the limit function h. A similar consideration, combined with the proof of 4.19, yields 4.31 (II). Theorem 4.31 is proved.

Now let G be an arbitrary domain $\subseteqq \mathbb{C}$, $G_R := \{t \in G : |t| < R\}$, $1 < q < 2$. We put

4.37 $$P_G h(z) = \lim_{R \to \infty} \left\{ P_{G_R} h(z) - \frac{1}{\pi} \iint_{G \setminus G_R} \frac{h(t)}{t - z} \, d\sigma_t \right\} \quad \text{for } h \in L_q(G),$$

where P_{G_R} is the operator defined in 4.30 for the bounded domain G_R. Note that the integral over $G \setminus G_R$ exists in any case as soon as $R > |z|$, and that, for these R, the expression behind the limit sign does not depend upon R. That is, for any bounded domain G' we have

4.38 $P_G h(z) = P_{G_R} h(z) - \dfrac{1}{\pi} \displaystyle\iint_{G \setminus G_R} \frac{h(t)}{t - z} \, d\sigma_t$ for almost all $z \in G'$ if R satisfies $G' \Subset \{|z| < R\}$.

4.39 ***Definition*** ([R3]). Let $h \in L_{p,q}(G)$, $1 < q < 2 < p < \infty$, $h = h_1 + h_2$ with $h_1 \in L_q(G)$, $h_2 \in L_p(G)$. We set

$$P_G^*(h_1 + h_2) := P_G h_1 + P_{0G} h_2(z),$$
$$P_G^* h = \{P_G^*(h_1 + h_2) : h_1 + h_2 = h,\ h_1 \in L_q(G),\ h_2 \in L_p(G)\}.$$

Instead of $P_{\mathbb{C}}^* h$, $P_{\mathbb{C}}^*(h_1 + h_2)$ we write only $P^* h$, $P^*(h_1 + h_2)$, respectively.

By this definition, for any $h \in L_{p,q}(G)$, $P_G^* h$ is a set of functions differing, however, only by additive constants. Indeed, if $h = h_1 + h_2 = h_1' + h_2'$ we have

4.40 $$P_G^*(h_1 + h_2)(z) - P_G^*(h_1' + h_2')(z) = P_G(h_1 - h_1')(z) - P_{0G}(h_2' - h_2)(z)$$
$$= P_G(h_1 - h_1')(0) \quad \text{where} \quad h_1 - h_1' = h_2' - h_2 \in L^{p,q}(G) \quad \text{(cf. 4.5)}.$$

Consequently, if any $f' \in P_G^* h$ has generalized derivatives $f_z', f_{\bar z}'$ in a domain G', then every $f \in P_G^* h$ has generalized derivatives $f_z, f_{\bar z}$ in G', and $f_z = f_z'$, $f_{\bar z} = f_{\bar z}'$ in G'. This fact leads to the following

4.41 ***Definition.*** We put

$$(P_G^* h)_z := f_z, \qquad (P_G^* h)_{\bar z} := f_{\bar z},$$

where f is any element of $P_G^* h$ possessing generalized derivatives $f_z, f_{\bar z}$ in G.

We then have the following statement.

4.42 *Every $f \in P_G^* h$ has generalized derivatives $f_z, f_{\bar z}$ in $\mathbb{C}$, namely*

$$(P_G^* h)_z = T_G h, \qquad (P_G^* h)_{\bar z} = h\omega_G.$$

For $f \in P_G^* h$ means that there are $h_1 \in L_q(G)$ and $h_2 \in L_p(G)$ with $h = h_1 + h_2$ and $f(z) = P_G^*(h_1 + h_2)(z) = P_G h_1(z) + P_{0G} h_2(z)$. Assertion 4.42 then follows from 4.37, 4.38, 4.31 (II) and 4.19 (II).

As the next theorem states, the existence of the derivative $f_{\bar z}$ implies the existence of f_z.

4.43 Theorem. *Let $f \in L_{p,q}(G)$, $1 < q < 2 < p < \infty$, and f have the generalized derivative $f_{\bar z} \in L_{p,q}(G)$ in the domain G. Then the generalized derivative f_z of f in G also exists and belongs to $L_{p,q}(G')$ for every bounded domain $G' \Subset G$. If $f \in L_r(G)$ and $f_{\bar z} \in L_r(G)$, $1 < r < \infty$, then $f_z \in L_{r,\mathrm{loc}}(G)$.*

Proof. Let $g \in P_G^*(f_{\bar z})$. Then $h := g - f$ belongs to $L_{1,\mathrm{loc}}(G)$ and has the generalized derivative $h_{\bar z} \equiv 0$ in G. By Weyl's lemma 3.13, h is equivalent to an analytic function in G and, thus, has a generalized derivative h_z in G. The theorem then follows from 4.42 and because T_G maps $L_r(G)$ as well as $L_{p,q}(G)$ into itself (concerning the case $r = 2$, note, e.g. 4.23 (III)).

Later we will occasionally need the following statement.

4.44 $$\iint_G f T_G g \, d\sigma = \iint_G g T_G f \, d\sigma \quad \forall f \in L_r(G),\ \forall g \in L_s(G), \quad r^{-1} + s^{-1} = 1.$$

Proof. Let $\varepsilon > 0$ and

4.45 $$T_{G\varepsilon}g(z) := -\frac{1}{\pi}\iint\limits_{|t-z|>\varepsilon} \frac{g(t)\,\omega_G(t)}{(t-z)^2}\,d\sigma_t = -\frac{1}{\pi}\iint\limits_G \frac{g(t)\,d_\varepsilon(t,z)}{(t-z)^2}\,d\sigma_t$$

where $d_\varepsilon(t, z) = 1$ if $|t - z| > \varepsilon$ and $= 0$ otherwise. By Fubini's theorem we have

$$\iint\limits_G fT_{G\varepsilon}g\,d\sigma = -\frac{1}{\pi}\iint\limits_G f(z)\left(\iint\limits_G \frac{g(t)\,d_\varepsilon(t,z)}{(t-z)^2}\,d\sigma_t\right)d\sigma_z$$

$$= -\frac{1}{\pi}\iint\limits_G g(t)\left(\iint\limits_G \frac{f(z)\,d_\varepsilon(t,z)}{(t-z)^2}\,d\sigma_z\right)d\sigma_t = \iint\limits_G gT_{G\varepsilon}f\,d\sigma.$$

Theorem 4.16 then yields

$$\iint\limits_G fT_G g\,d\sigma = \lim_{\varepsilon\to 0}\iint\limits_G fT_{G\varepsilon}g\,d\sigma = \lim_{\varepsilon\to 0}\iint\limits_G gT_{G\varepsilon}f\,d\sigma = \iint\limits_G gT_G f\,d\sigma,$$

which we had to prove.

Chapter I

QUASICONFORMAL MAPPINGS AND SOLUTIONS OF BELTRAMI SYSTEMS

Elliptic systems of partial differential equations of first order in the plane are closely connected with plane quasiconformal mappings. What are quasiconformal mappings? From the point of view of function theory one will describe them to be that generalization of conformal mappings which was introduced by H. Grötzsch in 1928. On the other hand, from the view point of partial differential equations, one will see them as homeomorphic solutions of Beltrami systems. This aspect of quasiconformal mappings is due to M. A. Lavrent'ev and C. B. Morrey. Both these meanings have melted into one notion, from 1928 onwards, during a period of about 35 years. In the course of this process two definitions have crystallized, the so-called geometric and the so-called analytic definition. We restrict our treatment of quasiconformal mappings to the analytic definition, which is sufficient for our purposes. The reader who wants to go beyond this frame may refer to [L/V] and [A2].

If the treatment of quasiconformal mappings is based upon the analytic definition then integral operators play an essential role, as we will see below. This approach has been developed in particular by I. N. Vekua, B. V. Bojarski and L. V. Ahlfors and L. Bers, cf. [V1], [B4], [A/B].

I.1 THE ANALYTIC DEFINITION OF QUASICONFORMAL MAPPINGS, BELTRAMI SYSTEMS

A mapping of a domain $G \subseteqq \overline{\mathbb{C}}$ onto a domain $G' \subseteqq \overline{\mathbb{C}}$ is said to be schlicht if it is, together with its inverse, single-valued, continuous and orientation-preserving.

1.1 ***Definition*** ([P1], cf. [L/V], pp. 168–170). Let $G \subseteq \overline{\mathbb{C}}$ be a domain, and K a real constant $\geqq 1$. The function $f(z)$ is called a K-quasiconformal mapping of G if and only if

(I) $f(z)$ is a schlicht mapping of G,

(II) $f(z) \in W^1_{1,\mathrm{loc}}(G \setminus \{\infty, f^{-1}(\infty)\})$ (f^{-1} is the inverse of f; if the point $a \in \overline{\mathbb{C}}$ is not contained in the domain of definition of the function g then we put $g(a) = \infty$ unless otherwise stated),

(III) $|f_{\bar z}| \leqq k|f_z|$ a.e. in G, $k = (K-1)/(K+1)$.

By 0.3.13 the 1-quasiconformal mappings are just the conformal mappings. (The condition 1.1 (III) could be replaced by

$$\text{1.1 (III}')\qquad D_f(z) := \frac{|f_z| + |f_{\bar z}|}{|f_z| - |f_{\bar z}|} \leqq K \quad \text{a.e. in } G,$$

cf. I.2.29. $D_f(z)$ is called the dilatation quotient of f at the point z, the conditions 1.1 (III), 1.1 (III′) are called dilatation conditions.)

To start with we prove the following

1.2 Theorem. *Every K-quasiconformal mappings $f(z)$ of a domain G is differentiable a.e. in G, belongs to $W^1_{2,\mathrm{loc}}(G \setminus \{\infty, f^{-1}(\infty)\})$ and, therefore, is absolutely continuous (as a homeomorphism of $G \setminus \{\infty, f^{-1}(\infty)\}$, cf. 0.2 above).*

Proof. Because of 0.2.11, 0.3.20 we only have to prove that f belongs to $W^1_{2,\mathrm{loc}}(G \setminus \{\infty, f^{-1}(\infty)\})$. Let G' be any domain compactly contained in $G \setminus \{\infty, f^{-1}(\infty)\}$. Then $\tau(e) := |f(e \cap G')|$ is a finite RBM, cf. 0.2. Since

$$|f_{\bar z}|^2 \leqq |f_z|^2 \leqq \frac{1}{1-k}(|f_z|^2 - |f_{\bar z}|^2) = \frac{1}{1-k}\tau'(z) \quad \text{a.e. in } G',$$

by 0.2.10, 0.2.11, and

$$\infty > \tau(G') \geqq \iint_{G'} \tau'(z)\, d\sigma_z,$$

by 0.2.7, we obtain $f_z, f_{\bar z} \in L_2(G')$, which we had to prove.

Let $\nu(z)$ be a measurable function defined a.e. in G such that

$$\textbf{1.3}\qquad |\nu(z)| \leqq k = \text{constant} < 1 \quad \text{a.e. in } G.$$

Such a $\nu(z)$ is called a Beltrami coefficient (in G), and the relation

$$\textbf{1.4}\qquad f_{\bar z}(z) = \nu(z) f_z(z),$$

which actually is a system of first order partial differential equations, cf. II.1 below, is called a Beltrami system (another terminology will be introduced in II.1).

1.5 ***Definition.*** The function $f(z)$ is called an L_p-solution of the Beltrami system 1.4 in the domain $G \subseteq \mathbb{C}$ if

(I) $f(z) \in W^1_{p,\mathrm{loc}}(G) \cap \mathscr{C}(G)$, $1 \leqq p < \infty$, and

(II) the generalized derivatives $f_z, f_{\bar{z}}$ satisfy 1.4 a.e. in G. An L_2-solution of 1.4 is simply called a solution of 1.4.

As a consequence of theorem 1.2 we have:

1.6 Theorem. (I) *Let $f(z)$ be a schlicht solution of the Beltrami system* 1.4 *in the domain $G \subseteq \mathbb{C}$. Then $f(z)$ is a quasiconformal mapping of G.*

(II) *Every quasiconformal mapping $f(z)$ of a domain G is a schlicht solution of the Beltrami system* 1.4 *in $G \setminus \{\infty, f^{-1}(\infty)\}$ with*

$$\nu(z) = \begin{cases} f_{\bar{z}}/f_z & \text{at every point of } G \text{ where } f_z, f_{\bar{z}} \text{ are defined and } f_z \neq 0, \\ 0 & \text{otherwise in } G. \end{cases}$$

Note that $f_z, f_{\bar{z}}$ are known to be defined a.e. in G, but so far we do not know whether f_z is different from zero a.e. in G.

Furthermore, the following lemma holds.

1.7 Lemma. (I) *Let $w(z)$ be any L_r-solution of the Beltrami system* 1.4 *in the domain $G \subseteq \mathbb{C}$, $r \geqq 1$. If $f(\cdot)$ is analytic in a domain $G_w \supseteq w(G)$, then $f \circ w(z)$ is again an L_r-solution of* 1.4 *in G.*

(II) *Let $w(z) \in W^1_{q,\mathrm{loc}}(G) \cap L_{s,\mathrm{loc}}(G)$, $1 < q < 2$, $s = q/(2-q)$, G as in* (I), *and let the generalized derivatives $w_z, w_{\bar{z}}$ satisfy* 1.4 *a.e. in G.*

If *there exists a schlicht mapping $w^*(z)$ of G having the properties*

(1) *$w^*(z)$ is an L_p-solution of* 1.4 *in G with $p^{-1} + q^{-1} = 1$,*

(2) *the inverse mapping $z(w^*)$ belongs to $W^1_{p,\mathrm{loc}}(w^*(G))$,*

then *there exists an analytic function f in $w^*(G)$ such that*

$$w(z) = f \circ w^*(z) \quad \text{a.e. in } G.$$

Proof. (I) is an immediate consequence of chain rule I, cf. 0.3.23 above. The proof of (II) runs as follows. Chain rule II in conjunction with theorem 0.3.20, applied to $f(w^*) := w \circ z(w^*)$, gives $f(w^*) \in W^1_{1,\mathrm{loc}}(w^*(G))$ and $f_{\bar{w}^*}(w^*) = 0$. Therefore, by Weyl's lemma, $w \circ z(w^*)$ is equivalent to an analytic function in $w^*(G)$. This proves 1.7 (II).

In what follows the solutions of Beltrami systems will shortly be called Beltrami solutions.

I.2 EXISTENCE AND PROPERTIES OF SCHLICHT BELTRAMI SOLUTIONS

In a certain manner, the statements in lemma 1.7 (II) embody a program leading to basic properties of Beltrami solutions and, simultaneously, of K-quasiconformal mappings. The crucial point is whether or not such mappings $w^*(z)$ mentioned in 1.7 (II) do exist. This question will be answered by the following

2.1 Theorem ([B4], [A/B], cf. also [M1], [B/N], [V1]). *Let $\nu(z)$ be a Beltrami coefficient in $\mathbb{C}$. There exists just one solution $f^\nu(z)$ of the corresponding Beltrami system* 1.4 *in $\mathbb{C}$ such that $f(0) = 0$, $f(1) = 1$ and*

(I) *$f^\nu(z)$ is a schlicht mapping of $\mathbb{C}$ onto $\mathbb{C}$,*

(II) *$f^\nu(z)$ and its inverse $(f^\nu)^{-1}$ belong to $W^1_{p,\mathrm{loc}}$ for every $p > 2$ satisfying $C(p) \cdot k < 1$ (cf. 1.3, 0.4.17, 0.4.18), and $|f_z|^2 - |f_{\bar z}|^2 > 0$ a.e. in $\mathbb{C}$.*

In proving this theorem we proceed as in [A2], chap. V A. Thus we first consider more special Beltrami coefficients, namely such of bounded support. Then we have the

2.2 Theorem ([B4], [A/B]). *Let $\nu(z)$ be a Beltrami coefficient in $\mathbb{C}$ of bounded support.*

(A) *There exists just one $f(z)$ with the properties*

(I) *f is an L_p-solution of the Beltrami system* 1.4 *in $\mathbb{C}$ with a $p > 2$,*

(II) *f is hydrodynamically normalized, that is, f has a Laurent expansion*

$$f(z) = z + \frac{a_1}{z} + \frac{a_2}{z^2} + \dots$$

in a neighbourhood of infinity.

(B) *This $f(z)$ has the properties mentioned in* 2.1(I), 2.1(II).

Proof of theorem 2.2. We start with proving uniqueness. Since diminution of $p > 2$ does not affect uniqueness we may choose p such that

2.3 $$2 < p, \quad kC(p) < 1,$$

cf. 0.4.17, 0.4.18. Further, because $f_{\bar z} = \nu f_z = 0$ in $\{|z| > R\}$ with a positive constant R, we have $f_{\bar z} \in L^{p,q}$ for every q subject to $1 \leqq q \leqq p$, especially for each fixed q satisfying

2.4 $$1 < q < 2, \quad kC(q) < 1.$$

The function $H(z) := f(z) - P(f_{\bar z})(z)$ is continuous in $\mathbb{C}$, and $H_{\bar z} = 0$ a.e. in $\mathbb{C}$. By Weyl's lemma $H(z)$ is analytic in $\mathbb{C}$. Moreover, by 2.2(II) and 0.4.3(IV), $H(z) - z$ tends to zero as $z \to \infty$, hence $H(z) = z$ and

2.5 $$f(z) = z + Ph(z), \qquad h = f_{\bar z} \in L^{p,q}.$$

On differentiating with respect to z and observing 1.4 and 0.4.19, we obtain the equation

2.6 $$h = \nu + \nu Th$$

for $h = f_{\bar z}$. Because of 2.3 and

2.7 $$\|\nu Th\|_{L_p} \leqq k\|Th\|_{L_p} \leqq kC(p)\,\|h\|_{L_p},$$

this equation has a unique solution by the Banach fixed point theorem. This proves the uniqueness of f.

In order to prove existence let h be a solution of 2.6 in L_p. This especially implies $h = 0$ in $\{|z| > R\}$, hence $h \in L^{p,q}$.

Consequently, $f(z) := z + Ph(z)$ belongs to $W^1_{p,\mathrm{loc}} \cap \mathscr{C}$, and this f is hydrodynamically normalized. Furthermore, by 0.4.19 we have

2.8 $$f_z = 1 + Th, \qquad f_{\bar z} = h.$$

This, together with 2.6, yields $f_{\bar z} = \nu f_z$ a.e. in $\mathbb{C}$. Assertion 2.2(A) is proved.

From 2.6 to 2.8 we further conclude

2.9 $\|f_{\bar z}\|_{L_p} \leqq (1 - kC(p))^{-1} \|\nu\|_{L_p}, \quad \|f_z - 1\|_{L_p} \leqq C(p)(1 - kC(p))^{-1} \|\nu\|_{L_p}$

and, because of 2.4, also

2.9′ $\|f_{\bar z}\|_{L_q} \leqq (1 - kC(q))^{-1} \|\nu\|_{L_q}, \quad \|f_z - 1\|_{L_q} \leqq C(q)(1 - kC(q))^{-1} \|\nu\|_{L_q}.$

The relations 2.5, 2.9, 2.9′ together with theorem 0.4.3 imply

2.10
$$|f(z_1) - f(z_2)| \leqq |z_1 - z_2|^{1-(2/p)} K_p(1 - kC(p))^{-1} \|\nu\|_{L_p} + |z_1 - z_2| \quad \forall z_1, z_2 \in \mathbb{C}$$

and

2.11
$$|z| - K_{p,q}(1 - kC(p, q))^{-1} \|\nu\|_{L^{p,q}} \leqq |f(z)| \leqq |z| + K_{p,q}(1 - kC(p, q))^{-1} \|\nu\|_{L^{p,q}},$$

where $C(p, q) = \max\{C(p), C(q)\}$.

Furthermore, the following lemma holds.

2.12 Lemma ([A2], cf. [V1]). *Let additionally $\nu \in W_p^1 \cap \mathscr{C}$ and $f(z)$ correspond to ν according to* 2.6, 2.5. *Then $f(z)$ is a continuously differentiable schlicht mapping of $\mathbb{C}$ onto itself, and $|f_z|^2 - |f_{\bar z}|^2 > 0$ everywhere in $\mathbb{C}$.*

Proof. If we were to differentiate both sides of the equation $f_{\bar z} = \nu f_z$ formally with respect to z we would obtain the equation

2.13
$$g_{\bar z} = (\nu g)_z = \nu g_z + \nu_z g$$

for $g = f_z$. Conversely, if g is a continuous function from $W^1_{p,\text{loc}}$ satisfying 2.13 a.e. in $\mathbb{C}$, then, by lemma 0.3.16, there exists a function $f^*(z) \in \mathscr{C}^1$ so that $f_z^* = g$, $f_{\bar z}^* = \nu g$ and, consequently, $f_{\bar z}^* = \nu f_z^*$.

For this reason we ask for a $g \in W^1_{p,\text{loc}} \cap \mathscr{C}$ satisfying 2.13. Since $\nu = \nu_z = 0$ for $|z| > R$ and because of 2.7, the equation

2.14
$$h = \nu T\nu_z + \nu Th$$

has a unique solution $h \in L_p$, which then must also belong to L_q, that is, $h \in L^{p,q}$. Then $P(h + \nu_z) =: \sigma \in W^1_{p,\text{loc}} \cap \mathscr{C}$ satisfies $\sigma_{\bar z} = \nu\sigma_z + \nu_z$ and, by chain rule I, $e^\sigma =: g \in W^1_{p,\text{loc}} \cap \mathscr{C}$ fulfils 2.13. Hence

$$f^*(z) = \int_0^z g\, dz + \nu g\, d\bar z$$

belongs to $\mathscr{C}^1$ and is a solution of 1.4. Moreover, $f_z^* = g \to e^0 = 1$ as $z \to \infty$ (cf. 0.4.3(IV)), thus

$$f^*(z) = z + a_0 + \frac{a_1}{z} + \frac{a_2}{z^2} + \ldots \quad \text{for } |z| > R.$$

As $|f_z^*|^2 - |f_{\bar z}^*|^2 = e^{2\operatorname{Re}\sigma}(1 - |\nu|^2) \geqq \text{constant} > 0 \ \forall z \in \mathbb{C}$, the restriction of $f^*(z)$ to sufficiently small disks is one-to-one. Therefore $f^*(\mathbb{C})$ is both closed and open in $\mathbb{C}$, hence $f^*(\mathbb{C}) = \mathbb{C}$. When following the pattern of the proof of the monodromy theorem (cf. [B2], p. 223) we then obtain that f^* is one-to-one in $\mathbb{C}$. Moreover, $f^*(z) - a_0$ has all properties mentioned in 2.2(I), (II).

By 2.2(A) it must then be equal to that $f(z)$ which corresponds to ν according to 2.6, 2.5. The lemma is proved.

The inverse function $f^{-1}(w)$ of this $f(z)$ also belongs to $\mathscr{C}^1$ and

2.15 $(f^{-1})_{\bar{w}} = \nu^{-1}(w)\,(f^{-1})_w$ where $\nu^{-1} := -\nu(f^{-1}(w))\,\overline{(f^{-1})_w}/(f^{-1})_w$,

cf. 0.3.27 above. Clearly,

2.16
$$|\nu^{-1}(w)| \leq k \quad \forall w \in \mathbb{C}, \qquad \nu^{-1}(w) = 0 \quad \text{for}$$
$$|w| > R^* := R + \frac{K_{p,q}}{1 - C(p,q)\,k}\,\|\nu\|_{L^{p,q}},$$

cf. 2.11. In particular, this again means that $\nu^{-1} \in L^{p,q}$. Of course, f^{-1} is again hydrodynamically normalized. Hence, by 2.2(A), f^{-1} is that function which belongs to the Beltrami coefficient ν^{-1} according to 2.6, 2.5. Consequently we also have

2.17
$$|z_1 - z_2| = |f^{-1}(w_1) - f^{-1}(w_2)|$$
$$\leq K_p(1 - kC(p))^{-1}\,\|\nu^{-1}\|_{L_p}\,|w_1 - w_2|^{1-(2/p)} + |w_1 - w_2|,$$

where $w_1 = f(z_1)$, $w_2 = f(z_2)$, cf. 2.10 above. For the L_p-norm of ν^{-1} we obtain

2.18
$$\|\nu^{-1}\|_{L_p} = (1 - kC(p))^{-2/p}\,\|\nu\|_{L_p},$$

because of

$$\iint |\nu^{-1}|^p\,d\sigma_w = \iint |\nu|^p(|f_z|^2 - |f_{\bar{z}}|^2)\,d\sigma_z \leq \iint |\nu|^p\,|f_z|^2\,d\sigma_z$$
$$= \iint |\nu|^{p-2}\,|f_{\bar{z}}|^2\,d\sigma_z \leq \|\nu\|_{L_p}^{p-2} \cdot \|f_{\bar{z}}\|_{L_p}^2$$

and 2.9.

Now let ν again be an arbitrary Beltrami coefficient in $\mathbb{C}$ with compact support, say, $\nu = 0$ for $|z| > R$, and let $f^{h,\nu}$ be that $f(z)$ which corresponds to ν according to 2.5, 2.6 (here h means hydrodynamical normalization). For each such ν there exists a sequence of ν_n, $n = 1, 2, \ldots$, such that

2.19 $|\nu_n(z)| \leq k$ $\forall z \in \mathbb{C}$, $\nu_n(z) = 0$ for $|z| > R + 1$ and every index n, $\|\nu_n - \nu\|_{L^{p,q}} \to 0$ and $\nu_n(z) \to \nu(z)$ pointwise a.e. in $\mathbb{C}$ as $n \to \infty$.

For instance, we may take $\nu_n(z) = \iint \nu(t)\,m_{\varrho_n}(t - z)\,d\sigma_t$ with a suitable sequence ϱ_n tending to zero as $n \to \infty$, cf. 0.3.8. Then we have, with $f_n := f^{h,\nu_n}$,

2.20
$$f_n(z) - f^{h,\nu}(z) \to 0 \quad \text{uniformly in } \mathbb{C} \text{ as } n \to \infty.$$

Indeed,
$$|f_n(z) - f^{h,\nu}(z)| = |Ph_n(z) - Ph(z)| \leq K_{p,q}\|h_n - h\|_{L^{p,q}},$$

cf. 0.4.3, where $h_n = \nu_n + \nu_n Th_n$ and $h = \nu + \nu Th$. Then $h_n - h$ is a solution of $h_n - h = (\nu_n - \nu)(1 + Th) + \nu_n T(h_n - h)$, hence

$$\|h_n - h\|_{L^{p,q}} \leq (1 - kC(p,q))^{-1}\,\|(\nu_n - \nu)(1 + Th)\|_{L^{p,q}} \to 0 \quad \text{as } n \to \infty.$$

This implies 2.20.

Furthermore, by 2.17 to 2.19 we have

2.21 $$|z_1 - z_2| \leqq K_p(1 - kC(p))^{-1-(2/p)} \|\nu\|_{L_p} \cdot |f_n(z_1) - f_n(z_2)|^{1-(2/p)} + |f_n(z_1) - f_n(z_2)| \quad \forall z_1, z_2 \in \mathbb{C},$$

and 2.21 remains valid if we replace f_n there by $f^{h,\nu}$, which follows by carrying out the limiting process $n \to \infty$ and using 2.20. Consequently, $f^{h,\nu}$ is one-to-one in $\mathbb{C}$. Then, by general results, cf. [L/V], p. 6, and when taking into account the hydrodynamical normalization, we see that $f^{h,\nu}$ is a homeomorphism of $\mathbb{C}$ onto itself. What we still have to prove is that, together with $f^{h,\nu}$, the inverse mapping $(f^{h,\nu})^{-1}$ also belongs to $W^1_{p,\mathrm{loc}}$. To this end we notice first that the inverse mapping f_n^{-1} of f_n corresponds to the Beltrami coefficient

$$\nu_n^{-1}(w) = -\nu_n(f_n^{-1}(w)) \overline{(f_n^{-1})_w}/(f_n^{-1})_w,$$

that is,

2.22 $$f_n^{-1}(w) = f^{h,\nu_n^{-1}}(w).$$

By 2.17 to 2.19 and 2.11, the functions f_n^{-1} are uniformly bounded and equicontinuous in bounded domains. Therefore, by the Ascoli–Arzelà theorem (cf. [Y], p. 85), a subsequence of the f_n^{-1}, without loss of generality the whole sequence, converges locally uniformly in $\mathbb{C}$ to a continuous function $\tilde{f}$. We also have

2.23 $$\begin{aligned} \|(f_n^{-1})_{\bar w}\|_{L_p} &\leqq (1 - kC(p))^{-1} \|\nu_n^{-1}\|_{L_p} \leqq (1 - kC(p))^{-1-(2/p)} \|\nu_n\|_{L_p} \\ &\leqq (1 - kC(p))^{-1-(2/p)} \|\nu\|_{L_p}, \\ \|(f_n^{-1})_w - 1\|_{L_p} &\leqq C(p)\,(1 - kC(p))^{-1-(2/p)} \|\nu\|_{L_p}. \end{aligned}$$

By 0.3.5 $\tilde{f}$ then belongs to $W^1_{p,\mathrm{loc}}$. But $\tilde{f} = (f^{h,\nu})^{-1}$. For since f_n^{-1} tends locally uniformly to $\tilde{f}$ we have

$$\tilde{f} \circ f^{h,\nu}(z) = \lim_{n\to\infty} \tilde{f} \circ f_n(z) = z + \lim_{n\to\infty} [\tilde{f}(f_n(z)) - f_n^{-1}(f_n(z))] = z \quad \forall z \in \mathbb{C},$$

and because of the local equicontinuity of the f_n in conjunction with locally uniform convergence of f_n^{-1} to f we obtain

$$f^{h,\nu} \circ \tilde{f}(w) = \lim_{n\to\infty} f_n \circ \tilde{f}(w) = w + \lim_{n\to\infty} [f_n(\tilde{f}(w)) - f_n(f_n^{-1}(w))] = w \quad \forall w \in \mathbb{C}.$$

From this it follows once more that $f^{h,\nu}$ is a homeomorphism of $\mathbb{C}$ onto $\mathbb{C}$ and, moreover, $(f^{h,\nu})^{-1} \in W^1_{p,\mathrm{loc}}$. Furthermore, by theorem 0.3.20, $f^{h,\nu}$ and $(f^{h,\nu})^{-1}$ are absolutely continuous homeomorphisms. Hence, by using the remarks underneath 0.2.12, we obtain $|f_z^{h,\nu}|^2 - |f_{\bar z}^{h,\nu}|^2 > 0$ a.e. in $\mathbb{C}$. By its hydrodynamical normalization $f^{h,\nu}$ is sense-preserving in a neighbourhood of infinity. Then, by a general result (cf. [L/V], p. 9) it is sense-preserving everywhere in $\mathbb{C}$. Assertion 2.2(B) is proved.

Proof of 2.1. Let ν be an arbitrary Beltrami coefficient in $\mathbb{C}$. The uniqueness of f^ν asserted in 2.1 is an immediate consequence of the fact that, by lemma 1.7(II), two such mappings may differ only by a Möbius transformation with fixed points 0, 1 and ∞, that is, by the identical mapping of $\mathbb{C}$ onto itself.

To prove the existence of such an f^ν we put

2.24 $\nu = \nu_1 + \nu_2$ where $\nu_1 = 0$ for $|z| \leqq 1$, $\nu_1 = \nu$ otherwise in $\mathbb{C}$, and $\nu_2 := \nu - \nu_1$. Then $\nu_0(z) := \nu_1\left(\frac{1}{z}\right)\frac{z^2}{\bar{z}^2}$ has a compact support $\subseteqq \{|z| \leqq 1\}$. Let $f_0(z) := f^{h,\nu_0}(z) - f^{h,\nu_0}(0)$. By chain rules 0.3.23, 0.3.29 and lemma 1.7(I)

2.25
$$W(z) := 1/f_0(1/z), \qquad W(0) = 0,$$

is a schlicht mapping of $\mathbb{C}$ onto itself, W, $W^{-1} \in W^1_{p,\text{loc}}$ (with the same p as in 2.3), and

2.26
$$W_{\bar{z}} = \nu_1 W_z, \qquad |W_z|^2 - |W_{\bar{z}}|^2 > 0 \quad \text{a.e. in } \mathbb{C}.$$

We now ask for a Beltrami coefficient $\lambda(\cdot)$ in $\mathbb{C}$ and a schlicht solution h of the corresponding Beltrami system $h_{\bar{w}} = \lambda(w)\, h_w$ with $h, h^{-1} \in W^1_{p,\text{loc}}$, so that $f := h \circ W$ is a schlicht L_1-solution (and, consequently, also an L_2-solution) of $f_{\bar{z}} = \nu f_z$ in $\mathbb{C}$. Suppose there is such a λ and such an h then by chain rule II it must hold

2.27
$$\nu(z) = \frac{(W_{\bar{z}}/\overline{W_z}) + \lambda \circ W(z)}{(W_z/\overline{W_z}) + \overline{\nu_1(z)}\, \lambda \circ W(z)}, \quad \text{hence } \lambda \circ W(z) = \frac{\nu_1(z) - \nu(z)}{1 - \nu(z)\,\overline{\nu_1(z)}}\,\frac{W_z}{\overline{W_z}}.$$

Conversely, because of the absolute continuity of $W(z)$ and $W_z \neq 0$ a.e. in $\mathbb{C}$ (cf. 2.26), $\lambda(w)$ can be defined a.e. in $\mathbb{C}$ by 2.27 and is then a measurable function satisfying $|\lambda(w)| \leqq 2k/(1+k^2) < 1$ a.e. in $\mathbb{C}$, $\lambda(w) = 0$ for $w \in W(\{|z| > 1\})$. Then $h(\cdot) := f^{h,\lambda}$ has all the required properties, and $f = h \circ W$ then indeed is a schlicht solution of $f_{\bar{z}} = \nu f_z$ in $\mathbb{C}$. By composing f with an entire linear transformation we obtain the desired f^ν, $f^\nu(0) = 0$, $f^\nu(1) = 1$. We still have to prove that $f^\nu, (f^\nu)^{-1} \in W^1_{p,\text{loc}}$. To this end let r be any positive number and $\nu_r = \nu$ if $|z| < r$, $\nu_r = 0$ otherwise. By theorem 2.2 and lemma 1.7(II) we have

2.28
$$f^\nu = F_r \circ f^{h,\nu_r} \quad \text{in } \{|z| < r\},$$

where F_r is analytic in $f^{h,\nu_r}(\{|z| < r\})$. Moreover, because of the schlichtness of f^ν and f^{h,ν_r}, F_r is conformal there. From this the remaining assertions follow. Theorem 2.1 is proved.

As a consequence of theorem 2.1, lemma 1.7(II) and chain rule I we obtain:

2.29 *Let $f(z)$ be a K-quasiconformal mapping of the domain $G \subseteqq \mathbb{C}$ into $\mathbb{C}$. Then*

(I) $|f_z|^2 - |f_{\bar{z}}|^2 > 0$, *especially* $f_z \neq 0$, *a.e. in* G,

(II) $f \in W^1_{p,\text{loc}}(G)$ *for each* $p > 2$ *for which* $\dfrac{K-1}{K+1}\,C(p) < 1$,

(III) f *is Hölder continuous in every domain* $G' \Subset G$.

Indeed let $\nu(z) = f_{\bar{z}}/f_z$ if $f_z \neq 0$, $z \in G$, and $\nu(z) = 0$ else. Then $f(z) = F \circ f^{\nu}(z)$, F analytic in $f^{\nu}(G)$. This proves (I) and (II). To prove (III) we put $\nu(z) = f_{\bar{z}}/f_z$ in G'', $G' \Subset G'' \Subset G$, and $\nu(z) = 0$ outside G''. Then $f(z) = H \circ f^{h,\nu}(z)$, H analytic in $f^{h,\nu}(G'')$. On combining this with 2.10 the assertion (III) follows.

Further let us notice:

2.30 *The inverse mapping of a K-quasiconformal mapping is a K-quasiconformal mapping, too.*

Obviously it is enough to prove 2.30 for K-quasiconformal mappings of bounded domains G onto bounded domains. Again let $\nu(z) = f_{\bar{z}}/f_z$ a.e. in G and $\nu = 0$ otherwise in $\mathbb{C}$. By lemma 1.7(II) we have $f = F \circ f^{h,\nu}$, F a conformal mapping of $f^{h,\nu}(G)$. Because of $(f^{h,\nu})^{-1} = f^{h,\nu^{-1}}$ with ν^{-1} specified in 2.15 we have $f^{-1} = f^{h,\nu^{-1}} \circ F^{-1}$. By 0.3.29 we obtain

$$(f^{-1})_{\bar{w}} = (f^{h,\nu^{-1}})_{\bar{\zeta}}\, \overline{(F^{-1})_w}, \quad (f^{-1})_w = (f^{h,\nu^{-1}})_{\zeta}\, (F^{-1})_w, \quad \text{where } \zeta = F^{-1}(w).$$

Hence

$$(f^{-1})_{\bar{w}}/(f^{-1})_w = \nu^{-1}(F^{-1}(w))\, \overline{(F^{-1})_w}/(F^{-1})_w, \qquad |\nu^{-1}(\zeta)| = |\nu(z)|$$

where

$$\zeta = F^{-1}(w) = f^{h,\nu}(z).$$

This proves 2.30.

2.31 *Let g be a K_1-quasiconformal mapping of G onto G_w and h be a K_2-quasiconformal mapping of G_w. Then $h \circ g$ is a K_1K_2-quasiconformal mapping of G.*

Proof. Clearly, $f := h \circ g$ is a schlicht mapping of G. Let G, G_w and $h(G_w)$ first be subsets of $\mathbb{C}$, in view of 1.1(II), (III). By chain rule II we have $f \in W^1_{1,\mathrm{loc}}(G)$, and with $\nu(z) := g_{\bar{z}}/g_z$, $\lambda(w) := h_{\bar{w}}/h_w$ it holds

$$\textbf{2.32} \quad |(h \circ g)_{\bar{z}}/(h \circ g)_z| = \left| \frac{(g_{\bar{z}}/\overline{g_z}) + \lambda(g(z))}{(g_z/\overline{g_z}) + \lambda(g(z))\,\overline{\nu(z)}} \right| \leqq \frac{|\nu| + |\lambda|}{1 + |\nu\lambda|} \leqq \frac{K_1K_2 - 1}{K_1K_2 + 1}$$

a.e. in G, where the inequalities in 2.32 hold on account of the elementary inequality

$$\left| \frac{a + b}{1 + \bar{a}b} \right| \leqq \frac{|a| + |b|}{1 + |ab|} \quad \forall a, b \text{ with } |a|, |b| < 1$$

and because of $|\nu| \leqq (K_1 - 1)/(K_1 + 1)$ a.e. in G, $|\lambda| \leqq (K_2 - 1)/(K_2 + 1)$ a.e. in G_w. Now let $G, G_w, h(G_w) \subseteqq \overline{\mathbb{C}}$. As just proved, f is a K_1K_2-quasiconformal mapping of $G \setminus \{\infty, g^{-1}(\infty), f^{-1}(\infty)\}$, especially $f \in W^1_{2,\mathrm{loc}}(G \setminus \{\infty, g^{-1}(\infty), f^{-1}(\infty)\})$, by theorem 1.2. Suppose, $g^{-1}(\infty) =: a \neq \infty$ is mapped under f onto $b \neq \infty$. By lemma 1.7(II) and theorem 2.1, f has the representation $f = F \circ W$ in $G \setminus \{\infty, a, f^{-1}(\infty)\}$, where $W(z)$ is a schlicht Beltrami solution in $\mathbb{C}$ and F is conformal in $W(G \setminus \{\infty\}) \setminus \{W(a), W(f^{-1}(\infty))\}$. By the theorem of Riemann (cf., e.g. [B/S], p. 114) $W(a)$ then is a removable singularity of F, that is, $f \in W^1_{2,\mathrm{loc}}(G \setminus \{\infty, f^{-1}(\infty)\})$. Assertion 2.31 is proved.

The same conclusion, by using the theorem of Riemann, also leads to the following statement.

2.33 *Let f be a K-quasiconformal mapping of G and a be an isolated boundary point of G. Then f can be extended to a K-quasiconformal mapping of $G \cup \{a\}$. In particular, any K-quasiconformal mapping of $\mathbb{C}$ onto $\mathbb{C}$ can by $f(\infty) = \infty$ be extended to a K-quasiconformal mapping of $\overline{\mathbb{C}}$ onto itself.*

By the way, the last assertion in 2.33 is also an immediate consequence of definition 1.1.

Furthermore, by means of lemma 1.7 and theorem 2.1, we can derive mapping theorems and statements on the boundary behaviour of quasiconformal mappings from corresponding statements on conformal mappings. For instance, by 1.7(I), 2.1 and Riemann's mapping theorem for conformal mappings of Jordan domains we at once obtain the

2.34 Theorem. *Let G be the interior domain of a (simply closed) Jordan curve $\mathfrak{C}$, and P_1, P_2, P_3 be three points on $\mathfrak{C}$ which follow upon another under positive orientation of $\mathfrak{C}$. For any Beltrami coefficient ν in G there exists just one schlicht mapping f of $G \cup \mathfrak{C}$ onto $\{|w| \leqq 1\}$, being a solution of $f_{\bar{z}} = \nu f_z$ in G and satisfying $f(P_1) = 1$, $f(P_2) = i$, $f(P_3) = -1$.*

I.3 COMPACTNESS AND CONVERGENCE PROPERTIES OF QUASICONFORMAL MAPPINGS

To start with we prove the

3.1 Lemma. (I) *Let $\mathfrak{N}(k, R)$ be the set of all Beltrami coefficients in $\mathbb{C}$ such that $\nu(z) = 0$ for $|z| > R$ and $|\nu| \leqq k < 1$ a.e. in $\mathbb{C}$. Then the $f^{h,\nu}$ with $\nu \in \mathfrak{N}(k, R)$ are locally equicontinuous, locally uniformly bounded and absolutely equicontinuous* (cf. 0.2).

(II) *Let $\nu_n \in \mathfrak{N}(k, R)$, $n = 1, 2, \ldots$ If the corresponding sequence $f_n := f^{h,\nu_n}$ is locally uniformly convergent, then the limit function f is a hydrodynamically normalized homeomorphism of $\mathbb{C}$ onto itself and $f, f^{-1} \in W^1_{p,\mathrm{loc}}$ for every p for which $kC(p) < 1$.*

Proof. The first two assertions in 3.1(I) follow at once from 2.10, 2.11, because the $\nu \in \mathfrak{N}(k, R)$ have uniformly bounded L_p- and L_q-norms. To prove the third assertion let W be any fixed bounded open set. For every set $e \subseteqq W$ there is a Borel set $e' \supseteqq e$ such that $|e'| = |e|$. Then we have

$$|f^{h,\nu}(e)| \leqq |f^{h,\nu}(e')| = \iint_{e'} (|f_z^{h,\nu}|^2 - |f_{\bar{z}}^{h,\nu}|^2)\, d\sigma_z \leqq \iint_{e'} |f_z^{h,\nu}|^2\, d\sigma_z$$

$$\leqq \|f_z^{h,\nu}\|_{L_p(W)}\, |e'|^{1-(2/p)} \leqq (\|f_z^{h,\nu} - 1\|_{L_p} + |W|^{1/p})\, |e|^{1-(2/p)},$$

cf. 0.3.20, 0.2.11, 0.2.12, 2.9. Assertion 3.1(I) is proved.

In order to prove 3.1(II) note first that the Beltrami coefficients ν_n^{-1} which correspond to the inverse mappings f_n^{-1} according to 2.15 belong to $\mathfrak{N}(k, 2R)$, by well-known estimates for hydrodynamically normalized conformal mappings of the domain $\{|z| > R\}$ (cf., e.g. [G1], p. 51). By 3.1(I) and the Ascoli–Arzelà theorem a subsequence of f_n^{-1} converges locally uniformly to a limit function $\tilde{f}$. As in the proof of theorem 2.2(B) we now conclude that $f, \tilde{f} \in W^1_{p,\mathrm{loc}}$ and $\tilde{f} = f^{-1}$ (cf. 2.23 and the subsequent considerations).

Of the convergence behaviour of quasiconformal mappings the following theorem has crucial importance.

3.2 Theorem ([S8]). *Let f_n be a sequence of K-quasiconformal mappings of the domain $G \subseteqq \mathbb{C}$ into $\mathbb{C}$ converging locally uniformly in G to a schlicht mapping f of G and satisfying*

$$|f_{n\bar{z}}/f_{nz}| =: k_n(z) \leqq k := \frac{K-1}{K+1} \quad \text{a.e. in } G \quad \forall n = 1, 2, \ldots$$

Then f is a K-quasiconformal mapping of G and

$$|f_{\bar{z}}/f_z| \leqq \limsup_{n\to\infty} k_n(z) \leqq k \quad \text{a.e. in } G.$$

Proof. Since we do not a priori assume the limit function f to be again a quasiconformal mapping (as is done in [S8] and which is, of course, assured by more general properties of sequences of quasiconformal mappings, cf. [L/V], p. 74) we must employ slightly different considerations (cf. [R1], p. 375) as in Strebel's original proof.

We need some lemmas.

3.3 Lemma. *Under the assumptions of theorem 3.2 the f_n are absolutely equicontinuous; the limit function f belongs to $W^1_{2,\mathrm{loc}}(G)$ and is therefore absolutely continuous, too.*

Proof. Let G', G'' be two domains such that $G' \Subset G'' \Subset G$. We put $\nu_n(z) = f_{n\bar{z}}/f_{nz}$ a.e. in G'', $\nu_n(z) = 0$ otherwise in $\mathbb{C}$. By 1.7(II) f_n has the representation

3.4 $\qquad f_n = F_n \circ g_n$ where $g_n := f^{h,\nu_n}$ and F_n is analytic in $g_n(G'')$.

By 3.1 the mappings g_n are locally equicontinuous and locally uniformly bounded, and the f_n, by hypothesis, are uniformly convergent in G''. Consequently, the F_n are uniformly bounded in $g_n(G'')$, and this implies uniform boundedness of the derivatives F_n' in $g_n(G')$, because we have $\mathrm{dist}(\partial g_n(G''), g_n(G')) \geqq \text{constant} > 0$ $\forall n$, on account of 3.1. This, in conjunction with 3.1(I), gives the first assertion. Furthermore, this implies that f belongs to $W^1_{p,\mathrm{loc}}(G)$ for every $p \geqq 2$ with $kC(p) < 1$. Indeed we have only to apply 0.3.5 and 2.9 to g_n. By theorem 0.3.20 f is then absolutely continuous. Lemma 3.3 is proved.

3.5 Lemma. *Let e be any measurable set $\Subset G$, and the conditions of theorem 3.2 be satisfied. Then*

$$\lim_{n\to\infty} |f_n(e)| = |f(e)|.$$

Proof. To start with let e_0 be any closed set $\Subset G$ and $E := f(e_0)$, $E_n := f_n(e_0)$, $n = 1, 2, \ldots$ Further, let $\omega_A(\cdot)$ be again the characteristic function of the arbitrary set $A \subseteq \mathbb{C}$. We have $\omega_E(w) = \omega_{e_0}(f^{-1}(w))$, $\omega_{E_n}(w) = \omega_{e_0}(f_n^{-1}(w))$. Since the f_n are uniformly convergent in e_0 we conclude:

3.6 *If $\omega_E(w) = 0$ then there is an n_0 such that $\omega_{E_n}(w) = 0 \quad \forall n \geqq n_0$.*

(This n_0 depends on $\operatorname{dist}(\{w\}, E)$.) Namely if $\omega_E(w) = 0$ then we have $\operatorname{dist}(\{w\}, E) = d > 0$, because E is closed. Suppose, there is a subsequence of f_n such that $f_n^{-1}(w) =: z_n \in e_0$. As $|f_n(z_n) - f(z_n)| < d/2$ if $n \geqq n_0(d)$, and $|f_n(z_n) - f(z_n)| = |w - f(z_n)| \geqq d$, we obtain a contradiction, which proves 3.6.

The statement 3.6 is equivalent to

3.7 $$\limsup_{n\to\infty} \omega_{E_n}(w) \leqq \omega_E(w) \quad \forall w \in \mathbb{C}.$$

Fatou's lemma (cf. [F1], p. 84) then gives $\limsup_{n\to\infty} |f_n(e_0)| \leqq |f(e_0)|$.

Now let e be any measurable set $\Subset G$. By lemma 3.3 for each $\varepsilon > 0$ there is a $\delta > 0$ so that $|f_n(e')| < \varepsilon$ for every $e' \subset e$ with $|e'| < \delta$. Moreover, for each $\delta > 0$ there is a closed set $e_0 \subset e$ such that $|e \setminus e_0| < \delta$ (by regularity of the Lebesgue measure). Hence

$$\limsup_{n\to\infty} |f_n(e)| = \limsup_{n\to\infty} (|f_n(e_0)| + |f_n(e \setminus e_0)|) \leqq |f(e_0)| + \varepsilon \leqq |f(e)| + \varepsilon,$$

and, since $\varepsilon > 0$ is arbitrary, we obtain

3.8 $$\limsup_{n\to\infty} |f_n(e)| \leqq |f(e)| \quad \text{for every measurable set } e \Subset G.$$

When starting with open sets instead of closed ones we obtain, in a corresponding way,

3.9 $$\liminf_{n\to\infty} |f_n(e)| \geqq |f(e)| \quad \text{for every measurable } e \Subset G.$$

Lemma 3.5 is proved. By the way, we only need 3.8 in what follows.

Let e be any measurable set $\Subset G$. We put

3.10 $$D(g, e) := \iint_e (|g_z|^2 + |g_{\bar z}|^2)\, d\sigma_z \quad \text{for any } g \in W^1_{2,\mathrm{loc}}(G).$$

3.11 Lemma (*lower semicontinuity of the Dirichlet integral*). *Under the assumptions of theorem 3.2 for any measurable set $e \Subset G$ it holds*

$$D(f, e) \leqq \liminf_{n\to\infty} D(f_n, e).$$

Indeed we have

3.12 $$0 \leqq D(f - f_n, e) = D(f, e) + D(f_n, e) - 2\operatorname{Re} \iint_e (f_{nz}\overline{f_z} + f_{n\bar z}\overline{f_{\bar z}})\, d\sigma.$$

After changing over to a subsequence we may assume that $D(f_n, e) \to \liminf_{n\to\infty} D(f_n, e) =: D$ as $n \to \infty$. By 0.3.5 we also have $f_{nz} \rightharpoonup f_z$, $f_{n\bar z} \rightharpoonup f_{\bar z}$ in $L_2(e)$. Letting n tend to infinity in 3.12, we obtain

$$0 \leqq D(f, e) + D - 2D(f, e),$$

which proves 3.11.

We put

3.13 $$k(z) := \limsup_{n\to\infty} k_n(z), \qquad M(z) := \frac{1 + k^2(z)}{1 - k^2(z)},$$

$$M_0 := \frac{1 + k^2}{1 - k^2} \quad (= \text{constant}).$$

3.14 Lemma. *Under the assumptions of theorem* 3.2, *for each measurable set* $e \Subset G$ *and each constant* M_e *satisfying* $M_e \geqq M(z)$ *a.e. in* e, *we have*

$$D(f, e) \leqq \limsup_{n\to\infty} D(f_n, e) \leqq M_e \cdot |f(e)|.$$

Indeed, let $m_n(z) = (1 + k_n^2(z))/(1 - k_n^2(z))$ and $M_n(z) = \sup\{m_j(z) : j \geqq n\}$. Then $\lim_{n\to\infty} M_n(z) = M(z)$. By 3.3 for each $\varepsilon > 0$ there is a $\delta(\varepsilon) > 0$ so that $|f_j(e \setminus e')| < \varepsilon\ \forall f_j$ and every $e' \subset e$ with $|e \setminus e'| < \delta(\varepsilon)$. By Egoroff's theorem 0.2.1 for each δ $(= \delta(\varepsilon))$ there is a closed set e' with $|e \setminus e'| < \delta$ whereon $M_n(z)$ tends uniformly to $M(z)$. Let M_e be any constant such that $M_e \geqq M(z)$ a.e. in e. Because of the uniform convergence of the $M_n(z)$ to $M(z)$ on e' we have $M_e + \varepsilon \geqq M_n(z)$ a.e. on e' $\forall n = n_0(\varepsilon)$. According to the definition of $k_n(z)$ and $m_j(z) \leqq M_n(z)$ $\forall j \geqq n$ this gives

$$D(f_j, e') \leqq (M_e + \varepsilon)\,|f_j(e')| \leqq (M_e + \varepsilon)\,|f_j(e)| \quad \forall j \geqq n_0(\varepsilon);$$

besides, we have

$$D(f_j, e) - D(f_j, e') = D(f_j, e \setminus e') \leqq M_0 |f_j(e \setminus e')| \leqq M_0 \varepsilon.$$

Hence

$$\limsup_{j\to\infty} D(f_j, e) \leqq \limsup_{j\to\infty} D(f_j, e') + M_0\varepsilon \leqq \limsup_{j\to\infty} (M_e + \varepsilon)\,|f_j(e)| + M_0\varepsilon$$
$$\leqq (M_e + \varepsilon)\,|f(e)| + M_0\varepsilon.$$

Since $\varepsilon > 0$ is arbitrary, this yields 3.14.

From 3.14 it now almost immediately follows the

Proof of 3.2. Suppose, $|f_{\bar z}| > k(z)\,|f_z|$ on a subset $G_0 \subset G$ of positive measure. Since G_0 is the union of the sets

$$G_n := \left\{z : |f_{\bar z}| > \left(k(z) + \frac{1}{n}\right)|f_z|\right\}, \qquad G_0 = \bigcup_{n=1}^{\infty} G_n,$$

there must exist at least one n so that $|G_n| > 0$. Then again, this G_n can be represented as the countable union of sets compactly contained in G, and at least one of them must also have positive measure. Hence there is an $\varepsilon > 0$ and a $G' \Subset G$ such that

$$|f_z|^2 + |f_{\bar z}|^2 \geqq (M(z) + \varepsilon)\,(|f_z|^2 - |f_{\bar z}|^2) \quad \text{in } G', \text{ and } |G'| > 0.$$

Furthermore, G' is the countable union of the sets

$$G''(M) := \left\{z : z \in G',\ M(z) \leqq M \leqq M(z) + \frac{\varepsilon}{2}\right\}, \quad \text{where } M \text{ is rational.}$$

Consequently, there is at least one rational number M such that $|G''(M)| > 0$. Thus

$$|f_z|^2 + |f_{\bar z}|^2 \geqq \left(M + \frac{\varepsilon}{2}\right)(|f_z|^2 - |f_{\bar z}|^2) \quad \forall z \in G''(M) \Subset G.$$

On integrating both sides over $e := G''(M)$ we obtain

3.15
$$D(f, e) \geqq \left(M + \frac{\varepsilon}{2}\right)|f(e)|,$$

where $|f(e)|$ is positive because of $|e| > 0$ and lemma 3.3. This is in contradiction with lemma 3.14. Therefore $|f_{\bar z}| \leqq k(z)\,|f_z|$ a.e. in G. Together with the schlichtness of f (by hypothesis) and lemma 3.3 this means that f is indeed a K-quasiconformal mapping of G. By 2.29(I) $f_z \neq 0$ a.e. in G, hence $|f_{\bar z}/f_z| \leqq k(z)$ a.e. in G, which is the last assertion in 3.2. Theorem 3.2 is proved.

3.16 Corollary. *If* $|f_{\bar z}/f_z| = \limsup_{n\to\infty} k_n(z)$ *in a set* $e \subset G$, $|e| > 0$, *then* $f_{nz} \to f_z$, $f_{n\bar z} \to f_{\bar z}$ *in* $L_2(e^*)$ *for every* $e^* \subseteqq e$, $e^* \Subset G$. *Consequently, there is a subsequence* $f_{n'}$ *such that* $f_{n'z} \to f_z$, $f_{n'\bar z} \to f_{\bar z}$ *pointwise a.e. in e.*

Since $f_z \neq 0$ a.e. in G, the last statement in 3.16 implies a result of K. Strebel [S8], namely that, under the assumptions of 3.16, there is a subsequence $f_{n'}$ such that

$$f_{n'\bar z}/f_{n'z} \to f_{\bar z}/f_z \quad \text{pointwise a.e. in } e.$$

Proof. Of course, the second assertion in 3.16 is a consequence of the first. To prove the first assertion let $e \Subset G$. For each $\delta > 0$ there is a closed set $e' \subset e$, $|e \setminus e'| < \delta$, so that the restriction of $M(z)$ to e' is uniformly continuous in e' (Lusin's theorem, cf., e.g. [F1], p. 76). Hence, for every $\varepsilon > 0$ there is a $d(\varepsilon) > 0$ so that

$$|M(z) - M(z')| < \varepsilon \quad \text{if } z, z' \in e' \text{ and } |z - z'| < d(\varepsilon).$$

Obviously, there is a subdivision of e' into a finite number of disjoint Borel sets $e'_j \subset e'$, $j = 1, \dots, N$, so that

$$e' = \bigcup_{j=1}^{N} e'_j \quad \text{and } \operatorname{diam} e'_j < d(\varepsilon), \quad j = 1, \dots, N,$$

for instance, cover e' by a suitable set of disjoint half-open squares S_j and put $e'_j = e' \cap S_j$. Then in each e'_j we have

3.17
$$(M(z_j) + \varepsilon)(|f_z|^2 - |f_{\bar z}|^2) \geqq |f_z|^2 + |f_{\bar z}|^2 \geqq (M(z_j) - \varepsilon)(|f_z|^2 - |f_{\bar z}|^2)$$

for any fixed $z_j \in e'_j$. In conjunction with 3.14 this gives

3.18
$$(M(z_j) - \varepsilon)\,|f(e'_j)| \leqq D(f, e'_j) \leqq \limsup_{n\to\infty} D(f_n, e'_j) \leqq (M(z_j) + \varepsilon)\,|f(e'_j)|,$$

hence

$$\limsup_{n\to\infty} D(f_n, e'_j) - D(f, e'_j) \leqq 2\varepsilon|f(e'_j)|, \quad j = 1, 2, \dots, N.$$

On summing up these inequalities and observing 3.11 we obtain

$$0 \leqq \limsup_{n\to\infty} D(f_n, e') - D(f, e') \leqq 2\varepsilon|f(e')|.$$

By 3.14 this further implies

$$0 \leqq \limsup_{n\to\infty} D(f_n, e) - D(f, e) \leqq 2\varepsilon|f(e')| + M_0|f(e \setminus e')|.$$

From this, and in conjunction with 3.12 and 0.3.5, we infer

3.19 $$\lim_{n\to\infty} D(f_n - f, e) = 0,$$

which gives the first assertion in 3.16. Corollary 3.16 is proved.

Let $\mathfrak{F}(k, R)$ be the set of all $f^{h,\nu}$ with $\nu \in \mathfrak{N}(k, R)$ (cf. 3.1 above). Further, let $\mathfrak{N}(k) := \{\nu : \nu \in L_\infty, |\nu| \leqq k$ a.e. in $\mathbb{C}\}$, and $\mathfrak{F}(k)$ be the set of all f^ν with $\nu \in \mathfrak{N}(k)$.

3.20 Theorem. (I) *$\mathfrak{F}(k, R)$ is compact (that means compact in itself) with respect to locally uniform convergence.*

(II) *The $f \in \mathfrak{F}(k)$ are locally uniformly bounded, $\mathfrak{F}(k)$ is locally Hölder continuous with exponent $1 - (2/p)$ for every $p > 2$ satisfying $kC(p) < 1$, and $\mathfrak{F}(k)$ is compact with respect to locally uniform convergence.*

Proof. (I) is a consequence of 3.1 (I), the Ascoli–Arzelà theorem, 3.1 (II) and theorem 3.2. In proving (II) let f_n be a sequence from $\mathfrak{F}(k)$. Each f_n can be represented by $f_n = h_n \circ W_n$ where W_n and h_n are to be constructed analogously to the mappings W and h in the proof of theorem 2.1 (see 2.25 to 2.27). By the statement (I), which we have just proved, and because of 3.1, 3.2, the mappings W_n are locally uniformly bounded, locally equicontinuous and relatively compact, with respect to locally uniform convergence, in the set of the $(1 + k)/(1 - k)$-quasiconformal mappings of $\mathbb{C}$ onto itself. The coefficients λ_n, which belong to W_n according to 2.27, have uniformly bounded supports, and $|\lambda_n| \leqq 2k/(1 + k^2)$ a.e. in $\mathbb{C}$ $\forall n$. Consequently, the mappings $h_n = f^{h,\lambda_n}$ are also locally uniformly bounded, locally equicontinuous and relatively compact in the set of $\left(\frac{1+k}{1-k}\right)$-quasiconformal mappings of $\mathbb{C}$ onto itself. Hence, the f_n are also locally uniformly bounded, locally equicontinuous, and every limit function of the f_n is a schlicht mapping. By the Ascoli–Arzelà theorem and theorem 3.2, $\mathfrak{F}(k)$, then, is compact with respect to locally uniform convergence. The local Hölder continuity of $\mathfrak{F}(k)$ with the exponents specified in 3.20 (II) follows, by means of the representations of the form 2.28, from 2.10 and the compactness of $\mathfrak{F}(k)$. Theorem 3.20 is proved.

3.21 Corollary. *Let $\mathfrak{M}$ be a set of K-quasiconformal mappings of $\mathbb{C}$ onto itself, $f(0) = 0$ $\forall f \in \mathfrak{M}$, and $\mathfrak{M}(a) := \{f(a) : f \in \mathfrak{M},\ a \in \mathbb{C} \setminus \{0\},\ a$ fixed$\}$ have neither 0 nor ∞ as accumulation points. Then $\mathfrak{M}$ is relatively compact in the set of the K-quasiconformal mappings of $\mathbb{C}$ onto itself.*

Proof. For each $f \in \mathfrak{M}$ we have a representation $f = F \circ f^{\nu}$ where $f^{\nu} \in \mathfrak{F}(k)$, $k = \dfrac{K-1}{K+1}$, and F is a conformal mapping of $\mathbb{C}$ onto itself with $F(0) = 0$. Hence, $f(z) = A_f f^{\nu}(z)$ where A_f is a constant $\neq 0$ for each $f \in \mathfrak{M}$. By the compactness of $\mathfrak{F}(k)$ and our assumption on $\mathfrak{M}(a)$, the set $\{A_f : f \in \mathfrak{M}\}$ cannot have 0 or ∞ as accumulation points, which gives our assertion. Corollary 3.21 is proved.

Now we place at our disposal, for later use, a lemma for estimating the absolute value of a function $w(z)$ in interior points of its domain of definition, by means of the L_q-norm of $w(z)$, if $w(z)$ has the form $f \circ \chi$ where f is analytic and χ quasiconformal.

3.22 Lemma. *Let $\chi(z)$ be a quasiconformal mapping of $\mathbb{C}$ onto itself, $|\chi_{\bar{z}}| \leqq k|\chi_z|$ a.e. in a bounded domain $G \subseteq \{|z| < R\}$, $k = \text{constant} < 1$. Outside G let χ be conformal and hydrodynamically normalized. Let the function f be analytical in $\chi(G)$, and let $w(z) := f \circ \chi(z)$ belong to $L_{q,\text{loc}}(G)$ for any fixed $q \in (1, 2)$ satisfying $kC\left(\dfrac{q}{q-1}\right) < 1$. Then there exists a constant $K(k, q, R)$ depending only upon k, q and R, so that*

$$|w(z)| \leqq K(k, q, R)\,[\text{dist}(\{z\}, \partial G')]^{-q/(2-q)} \cdot \|w\|_{L_q(G')} \quad \forall z \in G'$$

where G' is any open set $\Subset G$.

Proof. Let z_0 be any point $\in G'$, $\text{dist}(\{z_0\}, \partial G') = d > 0$. By the maximum principle for analytic functions and because χ is a homeomorphism, $\max\{|w(z)| : z \in G',\ \text{dist}(\{z\}, \partial G') \geqq d\}$ will be assumed at a point z_1 with $\text{dist}(\{z_1\}, \partial G') = d$. Thus

$$|w(z_1)| \geqq |w(z)| \quad \forall z \quad \text{satisfying} \quad \text{dist}(\{z\}, \partial G') \geqq d, \quad z \in G'.$$

In a neighbourhood of $\chi_1 := \chi(z_1)$ $f(\chi)$ has a Taylor expansion

$$f(\chi) = a_0 + a_1(\chi - \chi_1) + a_2(\chi - \chi_1)^2 + \dots,$$

and

$$|f \circ \chi(z)| \leqq |a_0| \quad \text{whenever} \quad \text{dist}(\{z\}, \partial G') \geqq d, \quad z \in G'.$$

Let $\alpha := (1 - (2/p))^{-1}$ with p to satisfy $p^{-1} + q^{-1} = 1$. Then

$$\textbf{3.23} \quad |a_0| \frac{d^{\alpha+1}}{\alpha+1} = \int_0^d |a_0|\, r^{\alpha}\, dr = \int_0^d \frac{1}{2\pi} \left| \oint_{\chi(\{|z-z_1|=r\})} \frac{f(\chi)}{\chi - \chi_1}\, d\chi \right| r^{\alpha}\, dr,$$

the latter by the Cauchy integral formula and by using the fact that $\chi(\{|z - z_1| = r\})$ is a rectifiable Jordan curve for almost all $r \in (0, d)$, cf. 0.3.10, 0.3.29. The last term is

$$\textbf{3.24} \quad \leqq \frac{1}{2\pi} \int_0^d r^{\alpha} \left(\int_0^{2\pi} \frac{|f \circ \chi(z)|}{|\chi(z) - \chi(z_1)|} |\chi_{\varphi}|\, d\varphi \right) dr \quad \text{where } z = z_1 + re^{i\varphi}.$$

By 2.17, 2.21 we have

3.25 $$\frac{|z - z_1|^\alpha}{|\chi(z) - \chi(z_1)|} \leqq K^*(k, q, R) \quad \text{everywhere in } G,$$

with a constant $K^*(k, q, R)$ depending only on q, k, R. Furthermore, by 0.3.10, 0.3.29 it holds

$$|\chi_\varphi| \leqq r(|\chi_x| + |\chi_y|) \leqq 2(|\chi_z| + |\chi_{\bar z}|)\,|z| \leqq 2(1 + k)\,|\chi_z| \cdot |z| \quad \text{a.e. in } \mathbb{C}.$$

Because of 3.25 and $\chi_z \in L_{p,\mathrm{loc}}$, we obtain

3.26 $$\frac{1}{2\pi} \iint\limits_{|z - z_1| < d} |f \circ \chi(z)| \frac{|z - z_1|^\alpha}{|\chi(z) - \chi(z_1)|} 2(1 + k)\,|\chi_z|\, d\sigma_z$$
$$\leqq K^*(k, q, R) \frac{1 + k}{\pi} \|w(z)\|_{L_q(G')} \|\chi_z\|_{L_p(G')}.$$

By Fubini's theorem the iterated integral in 3.24 then exists and is less than or equal to the right-hand term in 3.26. Thus

$$|a_0| \leqq (\alpha + 1) \frac{2}{\pi} K^*(k, q, R)\, \|\chi_z\|_{L_p(\{|z| < R\})}\, d^{-(\alpha+1)}\, \|w\|_{L_q(G')}.$$

By 2.9 this is our assertion. Lemma 3.22 is proved.

Chapter II

ELLIPTIC SYSTEMS OF PARTIAL DIFFERENTIAL EQUATIONS OF FIRST ORDER IN THE PLANE

In the first two sections of this chapter we mainly deal with the most general linear elliptic systems of first order partial differential equations in the plane and prove the Bers–Nirenberg representation theorem (which we formulate in a form suitable for our purposes). The representation theorem is both a central result on the solutions of elliptic systems and a basic tool for their treatment. We also turn to those more special systems which shall be the main topic of this book and which here will be called (ν, μ)-systems; their solutions we will call (ν, μ)-solutions.

In the third section we interpret (ν, μ)-systems, respectively (ν, μ)-solutions, in a geometric manner (cf. [V/L/A], problems 1395 to 1403, [L/S], p. 321). This is very useful, if not even inevitable, in order to make transparent certain involved estimates and calculations. In the fourth section we derive compactness results which will also be needed in later considerations, especially in chapter V and VI. The fifth section deals with differentiability properties of solutions of (ν, μ)-systems. In deriving differentiability our method rests on an idea of Bojarski ([B4], p. 499) and, in the long run, is an application of the principle which says that regularity problems in elliptic equations or systems may be treated by considering appropriate difference quotients of solutions which again prove to be solutions of elliptic equations or systems. This requires a certain condition on the coefficients, which is, roughly speaking, weaker than or at most equal to Hölder continuity and stronger than usual continuity. (That condition is possibly equivalent to Hölder continuity, at least in certain situations, but this question must here be left open.)

Subjects of the sixth section are adjoint systems and the (generalized) Cauchy integral theorem for (ν, μ)-solutions. As with analytic functions, with (ν, μ)-solutions the Cauchy integral theorem is the starting point for all further (contour-) integral theorems.

As to the treatment of more general elliptic systems of first order, especially in view of boundary value problems, we refer to [W] and the literature cited there.

II.1 CONDITIONS AND DEFINITIONS

Let the system of partial differential equations

1.1
$$\begin{aligned} v_x &= -cu_x - du_y + e_{11}u + e_{12}v + f_1, \\ v_y &= \quad au_x + bu_y + e_{21}u + e_{22}v + f_2 \end{aligned}$$

be given, with $a, b, c, d, e_{ij}, f_j, i, j = 1, 2$, to be real-valued measurable functions of the real variables x, y in a domain $G \subseteqq \mathbb{C}$. Such a system is called elliptic in G if

1.2
$$\Delta_0 := ad - (b + c)^2/4 > 0 \quad \text{and} \quad a > 0 \quad \text{a.e. in } G,$$

and uniformly elliptic if

1.3
$$\Delta_0 \geqq \text{constant} > 0, \ a > 0 \quad \text{a.e. in } G, \quad a, b, c, d \in L_\infty(G)$$

as well as

1.4
$$f_j, e_{ij} \in L^{p,q}(G), \quad 1 \leqq q < 2 < p < \infty.$$

When we put $w(z) = u(x, y) + iv(x, y)$, $z = x + iy$, and use the derivatives $w_z, w_{\bar{z}}$ (cf. 0.1.4, 0.3.3), the system 1.1 reads

1.5
$$w_{\bar{z}} = \nu w_z + \mu \overline{w_z} + Aw + B\bar{w} + F$$

where

1.6
$$\nu = \frac{d - a - i(b + c)}{N}, \qquad \mu = \frac{1 - \Delta + i(b - c)}{N},$$

$$\Delta = ad - bc, \qquad N = 1 + a + d + \Delta$$

and

1.7
$$A = \frac{-(e_{21} - ie_{22})(1 + d - ic) + i(e_{11} - ie_{12})(1 + a + ib)}{2N},$$
$$B = \frac{-(e_{21} + ie_{22})(1 + d - ic) + i(e_{11} + ie_{12})(1 + a + ib)}{2N},$$
$$F = \frac{-f_2(1 + d - ic) + if_1(1 + a + ib)}{N}.$$

By an easy calculation the condition 1.3 proves to be equivalent to

1.8
$$\| |\nu| + |\mu| \|_{L_\infty(G)} < 1.$$

If 1.3, respectively 1.8, is satisfied, then 1.4 holds if and only if

1.9
$$A, B, F \in L^{p,q}(G)$$

with the same p, q as in 1.4.

1.10 ***Definition.*** The function $w(z)$ is called an L_p-solution of the system 1.5 in the domain $G \subseteqq \mathbb{C}$ if

(I) $w(z) \in W^1_{p,\mathrm{loc}}(G) \cap \mathscr{C}(G)$, $1 \leqq p < \infty$,

(II) the generalized derivatives w_z, $w_{\bar{z}}$ satisfy 1.5 a.e. in G.

An L_2-solution is simply called a solution.

As already mentioned, our main subject here is the more special elliptic system

1.11
$$w_{\bar{z}} = \nu w_z + \mu \overline{w_z}, \qquad \| |\nu| + |\mu| \|_{L_\infty(G)} =: k < 1.$$

Such systems shall (also) be called (ν, μ)-systems, and their solutions (also) (ν, μ)-solutions. Then the $(\nu, 0)$-systems are just the Beltrami systems, and the $(0, \mu)$-systems correspond to the pseudoanalytic functions of the second kind in the sense of Bers (cf. [B1]) in the following way:

1.12 *Let μ be continuous in the domain G and $\|\mu\|_{L_\infty(G')} < 1$ for every subdomain $G' \Subset G$. Let $w(z)$ be a $(0, \mu)$-solution in every such G' and differentiable at every point of G. Then $w(z)$ is a pseudo-analytic function in G.*

By Polozii [P2] $(0, \mu)$-solutions are called (p, q)-analytic functions ($p = (1 - |\mu|^2)/|1 + \mu|^2$, $q = -2\,\mathrm{Im}\,\mu/|1 + \mu|^2$, cf. [P2], p. 10, as well as 1.13 below).

To every (ν, μ)-system 1.11 there corresponds a real system of partial differential equations 1.1 where $e_{ij} = f_j = 0$ for $i, j = 1, 2$, and

1.13
$$a = \frac{|1 - \nu|^2 - |\mu|^2}{\mathfrak{N}}, \qquad b = \frac{i(\nu - \bar{\nu} + \bar{\mu} - \mu)}{\mathfrak{N}},$$
$$\mathfrak{N} = |1 + \mu|^2 - |\nu|^2,$$
$$c = \frac{i(\nu - \bar{\nu} + \mu - \bar{\mu})}{\mathfrak{N}}, \qquad d = \frac{|1 + \nu|^2 - |\mu|^2}{\mathfrak{N}}$$
$$\Delta_0 = \frac{(1 + |\mu|^2 - |\nu|^2)^2 - 4|\mu|^2}{\mathfrak{N}^2}, \qquad \Delta = \frac{|1 - \mu|^2 - |\nu|^2}{\mathfrak{N}},$$

which can be shown by a simple (but perhaps a little troublesome) calculation.

Between elliptic systems 1.1, respectively 1.5, and elliptic partial differential equations of second order in the plane there exist close connections. Indeed, in the second order partial differential equation

1.14 $$A\varphi_{xx} + 2B\varphi_{xy} + C\varphi_{yy} + D\varphi_x + E\varphi_y + F = 0,$$

let A, B, C, D, E, F, say, be bounded and Hölder continuous in the domain G, and let equation 1.14 be uniformly elliptic in G, that is, there is a positive constant m such that

1.15 $$A(x, y)\,\xi^2 + 2B(x, y)\,\xi\eta + C(x, y)\,\eta^2 \geqq m(\xi^2 + \eta^2) \quad \forall (x, y) \in G, \ \forall \xi, \eta \in \mathbb{R}.$$

Further, let φ be any classical solution of 1.14. Then $w(z) = u(x, y) + iv(x, y) := \varphi_y(x, y) + i\varphi_x(x, y)$ is a solution of the uniformly elliptic system

1.16 $$v_x = -\frac{2B}{A}u_x - \frac{C}{A}u_y - Eu - Dv - F,$$
$$v_y = u_x.$$

Conversely, let a uniformly elliptic system 1.1 with Hölder continuously differentiable coefficients a, b, c, d, e_{ij}, f_j be given, and also let, for instance,

1.17 $$e_{12y} = e_{22x}.$$

By elimination of v_x, v_y we obtain from 1.1 the partial differential equation

1.18 $$(au_x + bu_y + e_{21}u + f_2)_x + (cu_x + du_y - e_{11}u - f_1)_y$$
$$+ e_{22}(-cu_x - du_y + e_{11}u + f_1) - e_{12}(au_x + bu_y + e_{21}u + f_2) = 0.$$

1.19 ***Remark.*** If $F = 0$ it is enough to require (also with $|G| = \infty$) that

(I) $A, B \in L_p(G)$, $2 < p < \infty$,

in place of 1.9.

Any system 1.5 with $F = 0$, satisfying 1.8 and 1.19 (I), shall be called uniformly elliptic, too.

II.2 THE REPRESENTATION THEOREM

The following form of the representation theorem (cf. [B/N], [B4], [V2], p. 174) is adequate for our purposes.

2.1 Representation theorem. *Let $w(z)$ be an L_q-solution of the system* 1.5 *whose coefficients ν, μ, A, B, F satisfy* 1.8, 1.9, *and let*

(1) $kC(q) < 1$, $1 < q < 2$, $k := \| |\nu| + |\mu| \|_{L_\infty(G)}$

as well as

(2) $kC(p) < 1$ *with p to satisfy* $p^{-1} + q^{-1} = 1$.

Then $w(z)$ is an L_p-solution of 1.5 *in G and admits the representation*

(I) $w(z) = e^{s(z)} f \circ \chi(z) + s_0(z) \quad \forall z \in G$.

Here $s(z)$, $s_0(z)$ *are bounded Hölder continuous functions in* $\mathbb{C}$ *belonging to* $W^1_{p,\mathrm{loc}}$. *The bounds, Hölder constants and Hölder exponents of* $s(z)$ *and* $s_0(z)$ *depend only on* k, p, q *and the* $L^{p,q}(G)$*-norms of* A, B *and* F; *moreover,*

$$\lim_{z\to\infty} s(z) = \lim_{z\to\infty} s_0(z) = 0 .$$

The function $\chi(\cdot)$ *is a* $(1+k)/(1-k)$*-quasiconformal mapping of* $\mathbb{C}$ *onto itself with* $\chi(0)=0$, $\chi(1)=1$, $\chi\in W^1_{p,\mathrm{loc}}$, *and* $f(\cdot)$ *is analytic in* $\chi(G)$.

If $F\equiv 0$ *in* G *then* $s_0(z)\equiv 0$ *in* $\mathbb{C}$. *If* $A=B\equiv 0$ *in* G *then* $s(z)\equiv 0$ *in* $\mathbb{C}$. *If* $\nu=\mu\equiv 0$ *in* G *then* $\chi(z)\equiv z$ *in* $\mathbb{C}$.

Proof. Obviously, $w(z)$ is an L_q-solution of 1.5 in G if and only if $w(z)$ is an L_q-solution of

2.2 $$w_{\bar z} = \tilde\nu w_z + \tilde A w + F$$

in G, where

2.3 $\tilde\nu = \nu + \mu \dfrac{\overline{w_z}}{w_z}$ $\quad \forall z\in G$ with $w_z \neq 0, \neq\infty$, $\quad \tilde\nu = 0$ elsewhere in $\mathbb{C}$

and

2.4 $\tilde A = A + B\dfrac{\bar w}{w}$ $\quad \forall z\in G$ with $w(z)\neq 0$, $\quad \tilde A = 0$ elsewhere in $\mathbb{C}$.

Then we have $\|\tilde\nu\|_{L_\infty} = k < 1$, $\tilde A(z)\in L^{p,q}$. Hence it is enough to prove theorem 2.1 for systems of the more special form 2.2. To this end we consider the equation

2.5 $$s_{\bar z} = \tilde\nu s_z + \tilde A \quad \text{in } \mathbb{C}.$$

A solution is (cf. 0.4.19)

2.6 $$s(z) = -\frac{1}{\pi}\iint \frac{h(t)+\tilde A(t)}{t-z}\, d\sigma_t$$

where h is the unique solution of the singular integral equation

2.7 $$h = \tilde\nu T\tilde A + \tilde\nu Th \quad \text{in } L^{p,q}.$$

By 0.4.3, 0.4.19 (cf. also I.2.7), this $s(z)$ has all the properties stated in 2.1. A direct computation, and applying 0.3.22, gives $W(z) := e^{-s(z)}w(z)$ to be an L_q-solution of

2.8 $$W_{\bar z} = \tilde\nu W_z + F e^{-s} \quad \text{in } G.$$

Now let $t(z)$ be that solution of the equation

2.9 $$t_{\bar z} = \tilde\nu t_z + F e^{-s} \quad \text{in } \mathbb{C}$$

which is constructed in the same way as $s(z)$ (that is, in 2.5 to 2.7 we have to replace $\tilde A$ by $F e^{-s}\in L^{p,q}$, $F=0$ outside G). This $t(z)$ has the same properties as $s(z)$. Further, $\tilde W := W - t$ is an L_q-solution of

2.10 $$\tilde W_{\bar z} = \tilde\nu \tilde W_z \quad \text{in } G.$$

Consequently, by I.1.7(II) and I.2.1, we have $\tilde{W} = f \circ \chi(z)$ where $\chi = f^{\tilde{\nu}}$ and f is analytic in $\chi(G)$. Thus $e^{-s(z)}w(z) = W(z) = \tilde{W}(z) + t(z)$ and

2.11 $$w(z) = e^{s(z)} f \circ \chi(z) + s_0(z) \quad \text{where} \quad s_0(z) := t(z)\, e^{s(z)}.$$

By 0.4.3, 2.7 we have the estimates

2.12 $$|s(z)| \leqq K_{p,q} \|h + \tilde{A}\|_{L^{p,q}} \leqq K_{p,q}(1 - kC(p, q))^{-1} (\|A\|_{L^{p,q}(G)} + \|B\|_{L^{p,q}(G)}),$$

where $C(p, q) := \max\{C(p), C(q)\}$,

2.13 $$|t(z)| \leqq e^{\max|s(z)|} K_{p,q}(1 - kC(p, q))^{-1} \|F\|_{L^{p,q}(G)},$$
$$|s_0(z)| \leqq \max|t(z)|\, e^{\max|s(z)|},$$

each for all $z \in \mathbb{C}$, and

2.14 $$|s(z_1) - s(z_2)| \leqq K_p(\|A\|_{L_p(G)} + \|B\|_{L_p(G)})\,(1 - kC(p))^{-1}\, |z_1 - z_2|^{1-(2/p)},$$

2.15 $$|t(z_1) - t(z_2)| \leqq K_p(1 - kC(p))^{-1}\, e^{\max|s(z)|} \|F\|_{L_p(G)}\, |z_1 - z_2|^{1-(2/p)}$$

as well as

2.16 $$|s_0(z_1) - s_0(z_2)| \leqq |t(z_1) - t(z_2)|\, e^{\max|s(z)|} + |e^{s(z_1)} - e^{s(z_2)}| \cdot \max|t(z)|$$

for any $z_1, z_2 \in \mathbb{C}$. The representation theorem is proved.

In particular we have the following.

2.17 Theorem. *Every (ν, μ)-solution $w(z)$ in the domain $G \subseteqq \mathbb{C}$ admits the representation*

(I) $w(z) = f \circ \chi(z)$

where χ is a K-quasiconformal mapping of $\mathbb{C}$ onto itself and f is analytic in $\chi(G)$. Hence, $w(z)$ is differentiable and $|w_z|^2 - |w_{\bar{z}}|^2 > 0$ a.e. in G, and every point z_0 of G, possibly up to a countable set of exceptional points, has a neighbourhood $U(z_0)$ such that the restriction of $w(z)$ to $U(z_0)$ is schlicht. If $w(z)$ is a schlicht L_1-solution of any (ν, μ)-system in a domain $G \subseteqq \mathbb{C}$, then $w(z)$ is a $(1 + k)/(1 - k)$-quasiconformal mapping of G with $k = \||\nu| + |\mu|\|_{L_\infty(G)}$.

All this is a consequence of theorem I.2.1 applied to $\chi = f^{\tilde{\nu}}$ ($\tilde{\nu}$ as in 2.3), and of the corresponding properties of analytic functions as well as of definition I.1.1.

In view of I.2.29 and chain rule II the following statement, though obvious by I.2.31, is a useful completion concerning the composition of (ν, μ)-solutions with quasiconformal mappings.

2.18 *Let $g(w) = f \circ \chi_1(w)$ where χ_1 is a K_1-quasiconformal mapping of the domain G_w and f is analytic in $\chi_1(G_w)$, and $\chi_2(z)$ be a K_2-quasiconformal mapping of G onto G_w. Then*

$$g \circ \chi_2(z) = f \circ \chi(z)$$

where χ is a K_1K_2-quasiconformal mapping of G.

Of course, the assumptions in the representation theorem 2.1 may be modified. For instance, in the homogeneous case $F \equiv 0$ (cf. [B4], [V2], p. 174) we also have the following

2.19 Representation theorem. *Let the functions $w(z)$, ν, μ as well as the constants p, q satisfy the conditions of theorem 2.1, A and B are to satisfy condition 1.19(I), and $F \equiv 0$. Then $w(z)$ is an L_p-solution of the homogeneous equation 1.5 and admits the representation*

(I) $w(z) = e^{s(z)} f \circ \chi(z) \quad \forall z \in G$

where f and χ have the same properties as in theorem 2.1, and $s(z)$ is defined in $\mathbb{C}$ and satisfies 2.14. If $A = B \equiv 0$ in G then $s(z) \equiv 0$. If $\nu = \mu \equiv 0$ in G then $\chi(z) \equiv z$.

The proof of 2.19 is essentially the same as that of 2.1, where at present $t(z) \equiv 0$, because $F \equiv 0$, and where we only have to use the P_0-operator in place of the P-operator for constructing a solution of 2.5, because $\tilde{A} \in L_p$, $p > 2$.

2.20 ***Remark.*** If ν and μ have bounded support in $\mathbb{C}$, then instead of $\chi = f^{\tilde{\nu}}$ the hydrodynamically normalized schlicht Beltrami solution $f^{h,\tilde{\nu}}$ may stand in 2.1(I) and 2.19(I).

From the representation theorem (in the one or other form) one can derive a lot of properties which the solutions of uniformly elliptic systems, and especially (ν, μ)-solutions, share with analytic functions. We come back to this fact in the fourth section of this chapter as well as in chapter III.

II.3 GEOMETRICAL INTERPRETATION OF (ν, μ)-SOLUTIONS

Let G be a domain $\subseteqq \mathbb{C}$, $w(z) = u(z) + iv(z)$ any (ν, μ)-solution in G and z_0 such a point of G where $w(z)$ is differentiable and its Jacobian

3.1 $$J_w(z_0) = |w_z(z_0)|^2 - |w_{\bar{z}}(z_0)|^2 > 0.$$

By

3.2 $$dw = w_z(z_0)\,dz + w_{\bar{z}}(z_0)\,d\bar{z} \quad (dw = du + i\,dv,\ dz = dx + i\,dy)$$

an affine mapping of the dz-plane onto the dw-plane is defined, which carries the ellipses in the dz-plane

3.3 $$\gamma\,dx^2 - 2\beta\,dx\,dy + \alpha\,dy^2 = \text{constant} =: h, \qquad \alpha\gamma - \beta^2 = 1,$$

into the ellipses

3.4 $$\gamma^*\,du^2 - 2\beta^*\,du\,dv + \alpha^*\,dv^2 = \text{constant} = h \cdot J_w(z_0)$$

in the dw-plane with

3.5 $$\gamma^* = (\gamma v_y^2 + 2\beta v_x v_y + \alpha v_x^2)/J_w(z_0),$$
$$\beta^* = (\gamma u_y v_y + \beta(u_x v_y + u_y v_x) + \alpha u_x v_x)/J_w(z_0),$$
$$\alpha^* = (\gamma u_y^2 + 2\beta u_x u_y + \alpha u_x^2)/J_w(z_0), \qquad \alpha^*\gamma^* - \beta^{*2} = 1.$$

As a terminology we will say in this case that the mapping $w(z)$ carries the infinitesimal ellipses 3.3 at the point z_0 (or the family of infinitesimal ellipses

3.3 with parameter h at the point z_0) into the infinitesimal ellipses 3.4 at the point $w_0 := w(z_0)$ (or into the family of infinitesimal ellipses 3.4 at the point w_0).

Now let any two families of infinitesimal ellipses

3.6 $S: \gamma\, dx^2 - 2\beta\, dx\, dy + \alpha\, dy^2 = \text{constant} = h$ and $S^*: \gamma^*\, du^2 - 2\beta^*\, du\, dv + \alpha^*\, dv^2 = \text{constant} = h^*$

be given, at z_0, where $\alpha\gamma - \beta^2 = \alpha^*\gamma^* - \beta^{*2} = 1$. The function $w(z)$ be differentiable and $J_w(z_0)$ be positive. What relations between the partial derivatives u_x, u_y, v_x, v_y at z_0 must hold in order that the mapping $w(z)$ carries the infinitesimal ellipses S at z_0 into the infinitesimal ellipses S^* at $w_0 = w(z_0)$? Of course, the relations 3.5 must hold, and the second of these gives

$$v_y(\gamma u_y + (\beta - \beta^*)\, u_x) + v_x((\beta + \beta^*)\, u_y + \alpha u_x) = 0,$$

thus

3.7 $$(\beta + \beta^*)\, u_y + \alpha u_x = \lambda v_y, \qquad \gamma u_y + (\beta - \beta^*)\, u_x = -\lambda v_x,$$

where λ is a still unknown constant, but which is different from zero, because $J_w(z_0) > 0$, in particular $(u_x, u_y) \neq (0, 0)$. From the third relation and the positive definiteness of the quadratic forms in 3.6 we infer $\lambda = \alpha^*$. Hence (cf. [V/L/A], problem 1403)

3.8 $$v_x = -\frac{\beta - \beta^*}{\alpha^*} u_x - \frac{\gamma}{\alpha^*} u_y,$$

$$v_y = \frac{\alpha}{\alpha^*} u_x + \frac{\beta + \beta^*}{\alpha^*} u_y,$$

and these are the relations between the partial derivatives which we were looking for.

Conversely, if $w(z)$ is differentiable at z_0 and satisfies 3.8, $\alpha\gamma - \beta^2 = \alpha^*\gamma^* - \beta^{*2} = 1$ and $J_w(z_0) > 0$, then $w(z)$ carries the infinitesimal ellipses S at z_0 into the infinitesimal ellipses S^* at $w(z_0)$.

Now let a linear uniformly elliptic system

3.9 $$v_x = -cu_x - du_y,$$

$$v_y = au_x + bu_y$$

in the domain G be given. With every point $z \in G$ we associate two families of ellipses

3.10 $$S(z): \gamma(z)\, dx^2 - 2\beta(z)\, dx\, dy + \alpha(z)\, dy^2 = h,$$

$$S^*(z): \gamma^*(z)\, du^2 - 2\beta^*(z)\, du\, dv + \alpha^*(z)\, dv^2 = h^*$$

such that

3.11 $$\frac{\alpha(z)}{\alpha^*(z)} = a(z), \qquad \frac{\beta(z) + \beta^*(z)}{\alpha^*(z)} = b(z),$$

$$\frac{\beta(z) - \beta^*(z)}{\alpha^*(z)} = c(z), \qquad \frac{\gamma(z)}{\alpha^*(z)} = d(z).$$

To this end, we have only to consider 3.11 to be a set of conditions which together with the requirement $\alpha\gamma - \beta^2 = \alpha^*\gamma^* - \beta^{*2} = 1$ determine $\alpha, \beta, \ldots, \gamma^*$ in a unique way. We then obtain

3.12 $$\alpha(z) = \frac{a(z)}{\sqrt{\Delta_0}}, \qquad \beta(z) = \frac{b(z) + c(z)}{2\sqrt{\Delta_0}}, \qquad \gamma(z) = \frac{d(z)}{\sqrt{\Delta_0}},$$
$$\alpha^*(z) = \frac{1}{\sqrt{\Delta_0}}, \qquad \beta^*(z) = \frac{b(z) - c(z)}{2\sqrt{\Delta_0}}, \qquad \gamma^*(z) = \frac{\Delta}{\sqrt{\Delta_0}}$$

(cf. [L/S], p. 321, [V/L/A], problem 1403); as to Δ_0, Δ see 1.2, 1.6. Together with 1.13 this gives

3.13 $$\alpha = (|1 - \nu|^2 - |\mu|^2)/\mathfrak{N}^*, \quad \beta = \mathrm{i}(\nu - \bar{\nu})/\mathfrak{N}^*, \quad \gamma = (|1 + \nu|^2 - |\mu|^2)/\mathfrak{N}^*,$$
$$\mathfrak{N}^* = ((1 - |\mu|^2 - |\nu|^2)^2 - 4|\nu\mu|^2)^{1/2} = ((1 + |\mu|^2 - |\nu|^2)^2 - 4|\mu|^2)^{1/2}$$
$$= ((1 + |\nu|^2 - |\mu|^2)^2 - 4|\nu|^2)^{1/2}$$

and

3.14 $$\alpha^* = (|1 + \mu|^2 - |\nu|^2)/\mathfrak{N}^* = \mathfrak{N}/\mathfrak{N}^*, \quad \beta^* = \mathrm{i}(\bar{\mu} - \mu)/\mathfrak{N}^*,$$
$$\gamma^* = (|1 - \mu|^2 - |\nu|^2)/\mathfrak{N}^*.$$

From 3.11, 3.12, 1.6 we obtain

3.15 $$\nu(z) = -\frac{\alpha - \gamma + 2\mathrm{i}\beta}{\alpha + \gamma + \alpha^* + \gamma^*}, \qquad \mu(z) = \frac{\alpha^* - \gamma^* + 2\mathrm{i}\beta^*}{\alpha + \gamma + \alpha^* + \gamma^*}.$$

When using the ratios p and p^* (each $\geqq 1$) of the major and minor semiaxes of the ellipses S and S^*, respectively, as well as the angles θ and θ^* between the positive x-axis and the major semiaxes of the ellipses S and S^*, respectively, we have

3.16
$$\nu(z) = -\frac{p - p^{-1}}{p + p^{-1} + p^* + p^{*-1}}\, \mathrm{e}^{2\mathrm{i}\theta}, \qquad \mu(z) = \frac{p^* - p^{*-1}}{p + p^{-1} + p^* + p^{*-1}}\, \mathrm{e}^{2\mathrm{i}\theta^*},$$
$$|\nu| + |\mu| \leqq (pp^* - 1)/(pp^* + 1)$$

(cf. [V/L/A], problem 1403).

Leaving aside the figurative background (or foreground) of the quadratic forms 3.10, we obtain the following statement as the quintessence of the above considerations.

3.17 Theorem. *Let the functions $u(z), v(z)$ have partial derivatives u_x, u_y, v_x, v_y, (in any sense) in the domain G, and let*

(I) $$A_z := \begin{pmatrix} \gamma(z) & -\beta(z) \\ -\beta(z) & \alpha(z) \end{pmatrix}, \qquad A_w := \begin{pmatrix} \gamma^*(z) & -\beta^*(z) \\ -\beta^*(z) & \alpha^*(z) \end{pmatrix}$$

be two positively definite matrices defined a.e. in G, $\det A_z = \det A_w = 1$ $\forall z \in G$ at which A_z, A_w are defined. Further, let z be any point of G with $J_w(z) = u_x v_y - u_y v_x > 0$ and A_z, A_w defined. Then the relations 3.8 hold at z

if and only if

$$\text{(II) } J_w(z)\, A_w = T_w^T A_z T_w \text{ where } T_w := \begin{pmatrix} v_y & -u_y \\ -v_x & u_x \end{pmatrix}.$$

(T_w^T means the transposed matrix of T_w.) 3.17(II) is equivalent to 3.5, and this again is equivalent to 3.8.

By means of this statement we can easily split up a (ν, μ)-solution into a chain of (ν_n, μ_n)-solutions. For this purpose we first derive a certain composition rule. Namely, let $w(z) = u(z) + iv(z)$, $\zeta(z) = \xi(z) + i\eta(z)$ be two functions defined in the domain G, and A_z, A_w, A_ζ be positively definite matrices defined a.e. in G, A_z, A_w as in 3.17 and

$$A_\zeta = \begin{pmatrix} \gamma_1(z) & -\beta_1(z) \\ -\beta_1(z) & \alpha_1(z) \end{pmatrix}, \qquad \det A_\zeta = 1.$$

The functions $w(z)$, $\zeta(z)$ are to be differentiable a.e. in G, their Jacobians $J_w(z)$, $J_\zeta(z)$ shall be positive a.e. in G, and the relation 3.17(II) as well as

3.18 $$J_\zeta(z)\, A_\zeta = T_\zeta^T A_z T_\zeta \quad \text{with} \quad T_\zeta = \begin{pmatrix} \eta_y & -\xi_y \\ -\eta_x & \xi_x \end{pmatrix}$$

are to hold a.e. in G. This means by 3.17 that we have

3.19 $$\eta_x = -\frac{\beta - \beta_1}{\alpha_1}\xi_x - \frac{\gamma}{\alpha_1}\xi_y, \qquad v_x = -\frac{\beta - \beta^*}{\alpha^*}u_x - \frac{\gamma}{\alpha^*}u_y,$$

$$\eta_y = \frac{\alpha}{\alpha_1}\xi_x + \frac{\beta + \beta_1}{\alpha_1}\xi_y, \qquad v_y = \frac{\alpha}{\alpha^*}u_x + \frac{\beta + \beta^*}{\alpha^*}u_y \quad \text{a.e. in } G.$$

If, in addition, $\zeta(z)$ is an absolutely continuous homeomorphism of G, then its inverse $z(\zeta)$ is also differentiable a.e. in $\zeta(G)$. Then the composition $w^*(\zeta) := w \circ z(\zeta) =: u^*(\zeta) + iv^*(\zeta)$ is also differentiable a.e. in $\zeta(G)$, and for the partial derivatives of w^* with respect to $\zeta, \bar{\zeta}$ the usual chain rule holds.

Our composition rule now reads

3.20 $$v_\xi^* = -\frac{\beta_1 - \beta^*}{\alpha^*}u_\xi^* - \frac{\gamma_1}{\alpha^*}u_\eta^*$$

$$v_\eta^* = \frac{\alpha_1}{\alpha^*}u_\xi^* + \frac{\beta_1 + \beta^*}{\alpha^*}u_\eta^* \quad \text{a.e. in } \zeta(G),$$

where $\alpha_1 = \alpha_1(z(\zeta)), \ldots, \gamma^* = \gamma^*(z(\zeta))$. The proof is now simple: By 3.17(II), 3.18 $J_w(z)\, A_w = J_\zeta(z)\, T_w^T((T_\zeta^T)^{-1} A_\zeta T_\zeta^{-1})\, T_w$, and, because $J_w(z)/J_\zeta(z) = J_{w^*}(\zeta)$, this implies

$$J_{w^*}(\zeta)\, A_w = T_{w^*}^T A_\zeta T_{w^*} \quad \text{with} \quad T_{w^*} = T_\zeta^{-1} T_w = \begin{pmatrix} v_\eta^* & -u_\eta^* \\ -v_\xi^* & u_\xi^* \end{pmatrix},$$

which is equivalent to 3.20, again by 3.17.

This composition rule, which is a kind of transitivity law, could be formalized as follows:

3.21 If $(\zeta, A_\zeta) \curvearrowright (z, A_z)$ and $(z, A_z) \curvearrowright (w, A_w)$, then $(\zeta, A_\zeta) \curvearrowright (w, A_w)$.

Here $(z, A_z) \curvearrowright (w, A_w)$ means the relation 3.17(II), and $(\zeta, A_\zeta) \curvearrowleft (z, A_z)$ means 3.17(II) with respect to $\zeta(z)$, i.e. 3.18, together with the existence of a sufficiently well-behaved inverse $z(\zeta)$.

The ellipses $S(z)$ and $S^*(z)$ in 3.10 shall also be called A_z-ellipses and A_w-ellipses, respectively, if A_z and A_w are given by 3.17(I) with the same $\alpha, \beta, \gamma, \ldots, \gamma^*$ as in 3.10. We will say that the A_z-ellipses (respectively the A_w-ellipses) correspond to $\nu(z), \mu(z)$ if α, β, γ, (respectively $\alpha^*, \beta^*, \gamma^*$) are defined by 3.13 (respectively 3.14).

As an example we want to split up a (ν, μ)-solution $w(z)$ defined in a domain $G \subseteqq \mathbb{C}$ into two functions $\zeta(z)$, $w^*(\zeta)$, where $\zeta(z)$ is to be a schlicht mapping carrying the infinitesimal A_z-ellipses corresponding to ν, μ at almost every point $z \in G$ into infinitesimal circles, i.e.

$$A_\zeta = \begin{pmatrix} 1 & 0 \\ 0 & 1 \end{pmatrix},$$

and $w^*(\zeta)$ is to carry infinitesimal circles at almost every point ζ of $\zeta(G)$ into the infinitesimal A_w-ellipses corresponding to $\nu = \nu(z(\zeta))$, $\mu = \mu(z(\zeta))$.

Accordingly, by 3.12 and 3.17

3.22 $$A_z = \frac{1}{\sqrt{\Delta_0}} \begin{pmatrix} d & -(b+c)/2 \\ -(b+c)/2 & a \end{pmatrix}, \qquad A_\zeta = \begin{pmatrix} 1 & 0 \\ 0 & 1 \end{pmatrix}.$$

According to 3.11, 3.17 (with A_w replaced by A_ζ) $\zeta(z) = \xi(z) + i\eta(z)$ has to satisfy the Beltrami system

3.23 $$\eta_x = -\frac{b+c}{2\sqrt{\Delta_0}}\,\xi_x - \frac{d}{\sqrt{\Delta_0}}\,\xi_y,$$

$$\eta_y = \frac{a}{\sqrt{\Delta_0}}\,\xi_x + \frac{b+c}{2\sqrt{\Delta_0}}\,\xi_y,$$

or, in complex notation,

3.24 $$\zeta_{\bar z} = \nu^* \zeta_z, \qquad \nu^* = \frac{2\nu}{\mathfrak{N}^* + 1 + |\nu|^2 - |\mu|^2},$$

cf. 3.13, 3.15. With $w^*(\zeta) = u^*(\zeta) + iv^*(\zeta)$ we have to associate the matrices

3.25 $$A_\zeta = \begin{pmatrix} 1 & 0 \\ 0 & 1 \end{pmatrix}, \qquad A_w = \frac{1}{\sqrt{\Delta_0}} \begin{pmatrix} \Delta & -(b-c)/2 \\ -(b-c)/2 & 1 \end{pmatrix},$$

where $\Delta_0 = \Delta_0(z(\zeta))$, $\Delta = \Delta(z(\zeta))$, $b = b(z(\zeta))$, $c = c(z(\zeta))$, and $z(\zeta)$ is the inverse of $\zeta(z)$. This means that (according to 3.11, 3.17) $w^*(\zeta)$ is to satisfy

3.26 $$v^*_\xi = \frac{b-c}{2}\,u^*_\xi - \sqrt{\Delta_0}\,u^*_\eta,$$

$$v^*_\eta = \sqrt{\Delta_0}\,u^*_\xi + \frac{b-c}{2}\,u^*_\eta,$$

or, in complex notation,

3.27

$$w^*_{\bar\zeta} = \mu^* \overline{w^*_\zeta}, \qquad \mu^* = \frac{2\mu}{\mathfrak{N}^* + 1 + |\mu|^2 - |\nu|^2}, \qquad \nu = \nu(z(\zeta)), \qquad \mu = \mu(z(\zeta)).$$

Hence, let $\zeta(z)$ be a schlicht solution of 3.24 in the domain G, and $w^*(\zeta)$ be a $(0, \mu^*)$-solution in the domain $\zeta(G)$. Then $w^* \circ \zeta(z)$ is a (ν, μ)-solution in G.

Conversely, $w^*(\zeta) := w \circ z(\zeta)$ is a solution of 3.27 if $w(z)$ is a (ν, μ)-solution in G and $\zeta(z)$ is a schlicht solution of 3.24 in G. Note that ν^* in 3.24 depends only on ν and μ (but not on $w(z)$), μ^* in 3.27 depends only on ν, μ and $\zeta(z)$ (but not on $w(z)$, either). This means that we can reduce (ν, μ)-systems to $(0, \mu^*)$-systems by means of schlicht Beltrami solutions (but this reduction does not retain certain properties, so that it is inevitable for certain questions to consider the general (ν, μ)-systems).

In place of

$$A_\zeta = \begin{pmatrix} 1 & 0 \\ 0 & 1 \end{pmatrix},$$

which led to splitting up a (ν, μ)-solution into a Beltrami solution and a $(0, \mu^*)$-solution, we could have used any other positively definite symmetric $(2, 2)$-matrix A with $\det A = 1$, which would have led to splitting up a (ν, μ)-solution into a schlicht (ν_1, μ_1)-solution $\zeta(z)$ (where ν_1, μ_1 depend only on ν, μ and A) and a (ν_2, μ_2)-solution $w^*(\zeta)$ (where ν_2, μ_2 depend only on ν, μ, A and $\zeta(z)$). In the same way we can split up any (ν, μ)-solution into a chain of an arbitrary number of (ν_n, μ_n)-solutions.

For later use we want to collect some elementary formulae about the behaviour of (ν, μ)-systems under affine mappings. Thus, let $w(z)$ be a (ν, μ)-solution in the domain G and

3.28 $$\zeta = z + \ell \bar z + d, \qquad \ell, d = \text{constant}, \quad |\ell| < 1,$$

an affine mapping of the z-plane. We put $w^*(\zeta) = w \circ z(\zeta)$. While $w(z)$ carries the S-ellipses in 3.6 into the S^*-ellipses in 3.6, $w^*(\zeta)$ carries certain S_1-ellipses into those S^*-ellipses. Here the S_1-ellipses are the images of the S-ellipses under the affine mapping 3.28, by 3.17 to 3.21. Hence 3.20 holds a.e. in G. The parameters $\alpha_1, \beta_1, \gamma_1$ of the S_1-ellipses are determined by the α, β, γ of the S-ellipses and by 3.28 and can easily be calculated. According to 3.15, by means of $\alpha_1, \beta_1, \gamma_1$ we can write down a system of partial differential equations for $w^*(\zeta)$. On the other hand, from 3.28 we can also directly calculate a new system for $w^*(\zeta)$. In both cases one obtains the same, namely

3.29 $$w^*_{\bar\zeta} = \nu_1(\zeta)\, w^*_\zeta + \mu_1(\zeta)\, \overline{w^*_\zeta}$$

with

3.30 $$\nu_1(\zeta) = (\bar\nu \ell^2 - (1 + |\nu|^2 - |\mu|^2)\, \ell + \nu)/\mathfrak{N}_1, \qquad \mu_1(\zeta) = \mu(1 - |\ell|^2)/\mathfrak{N}_1,$$
$$\mathfrak{N}_1 = |1 - \bar\nu \ell|^2 - |\mu \ell|^2, \qquad \nu = \nu(z(\zeta)), \qquad \mu = \mu(z(\zeta)).$$

In particular, $\nu_1(\zeta)$ vanishes at a prescribed point $\zeta_0 = z_0 + \ell\overline{z_0} + d \in G_\zeta$, where G_ζ is the image domain of G under the mapping 3.28, if we choose

3.31 $\ell = 2\sigma/(1 + \sqrt{1 - 4|\sigma|^2}), \qquad \sigma = \nu(z_0)/(1 + |\nu(z_0)|^2 - |\mu(z_0)|^2).$

In keeping with 3.24 we have

3.32

$$\begin{aligned}\ell &= 2\nu(z_0)/[1 + |\nu(z_0)|^2 - |\mu(z_0)|^2 + \sqrt{(1 + |\nu(z_0)|^2 - |\mu(z_0)|^2)^2 - 4|\nu(z_0)|^2}] \\ &= \nu(z_0) + 2\overline{\mu(z_0)} \cdot \tau/(1 + \sqrt{1 - 4|\tau|^2}), \\ &\qquad \tau = \nu(z_0)\,\mu(z_0)/[1 - |\nu(z_0)|^2 - |\mu(z_0)|^2], \\ &\qquad\qquad |\tau| \leqq k^2/2 \text{ (if } |\nu(z_0)| + |\mu(z_0)| \leqq k) \text{ and (then) } |\ell| \leqq k.\end{aligned}$$

Let $w(z)$ again be a (ν, μ)-solution in the domain G. We now subject the w-plane to an affine mapping

3.33 $$h = w + b\bar{w}, \qquad b = \text{constant}, \quad |b| < 1.$$

As an easy calculation shows, the function $h(z) := w(z) + b\overline{w(z)}$ is a (ν_0, μ_0)-solution in G with

3.34 $$\nu_0(z) = \nu \cdot (1 - |b|^2)/\mathfrak{N}_0, \quad \mu_0(z) = (\bar{\mu}b^2 + (1 + |\mu|^2 - |\nu|^2)\, b + \mu)/\mathfrak{N}_0,$$
$$\mathfrak{N}_0 = |1 + \bar{\mu}b|^2 - |\nu b|^2, \quad \nu = \nu(z), \quad \mu = \mu(z).$$

This means geometrically that the S-ellipses remain unchanged and the S^*-ellipses are subjected to the affine mapping 3.33.

We obtain $\mu_0(z_0) = 0$ at $z_0 \in G$ if we put

3.35 $b = 2\varkappa/(1 + \sqrt{1 - 4|\varkappa|^2}), \qquad \varkappa = -\mu(z_0)/(1 + |\mu(z_0)|^2 - |\nu(z_0)|^2).$

On comparing this with 3.31, 3.32 we see that, with the same τ as there,

3.36 $$b = -\mu(z_0) - 2\overline{\nu(z_0)}\,\tau/(1 + \sqrt{1 - 4|\tau|^2}), \quad |b| \leqq k.$$

We now put $g(\zeta) := w(z(\zeta)) + b\overline{w(z(\zeta))}$ where $z(\zeta)$ is the affine mapping defined by 3.28 and b is any constant as in 3.33. For this $g(\zeta)$ we can establish a (ν_2, μ_2)-system by applying the affine mapping 3.28 to the (ν_0, μ_0)-system for $h(z)$ and calculating ν_2, μ_2 in the same way from ν_0, μ_0 as ν_1, μ_1 in 3.30 from ν, μ. On the other hand, we could start with the (ν_1, μ_1)-system 3.29, 3.30 for $w^*(\zeta) = w(z(\zeta))$ and then apply the affine mapping 3.33, where we had to calculate the coefficients ν_2^*, μ_2^* of the new system from ν_1, μ_1 as ν_0, μ_0 in 3.34 from ν, μ. It is not surprising, and also not trivial, that $\nu_2 = \nu_2^*$, $\mu_2 = \mu_2^*$,

3.37
$$\begin{aligned}\nu_2(\zeta) &= [\bar{\nu}\ell^2 - (1 + |\nu|^2 - |\mu|^2)\,\ell + \nu]\,(1 - |b|^2)/\mathfrak{N}_2, \\ \mu_2(\zeta) &= [\bar{\mu}b^2 + (1 + |\mu|^2 - |\nu|^2)\,b + \mu]\,(1 - |\ell|^2)/\mathfrak{N}_2, \\ \mathfrak{N}_2 &= (1 - |b|^2)\,[\tfrac{1}{2}(1 + |\nu|^2 - |\mu|^2)\,(1 + |\ell|^2) - 2\,\mathrm{Re}\,\ell\bar{\nu}] \\ &\quad + (1 - |\ell|^2)\,[\tfrac{1}{2}(1 + |\mu|^2 - |\nu|^2)\,(1 + |b|^2) - 2\,\mathrm{Re}\,b\bar{\mu}], \\ &\qquad \nu = \nu(z(\zeta)), \qquad \mu = \mu(z(\zeta)).\end{aligned}$$

Besides this, for ℓ and b from 3.31 and 3.35, respectively, we have

3.38 $$\ell = \nu(z_0) + 2\overline{\mu(z_0)} \cdot \tau/(1 + \sqrt{1 - 4|\tau|^2}) = \nu_0(z_0),$$
$$b = -\mu(z_0) - 2\overline{\nu(z_0)} \cdot \tau/(1 + \sqrt{1 - 4|\tau|^2}) = -\mu_1(\zeta_0),$$

where ν_0 is to form according to 3.34 with b from 3.36, and μ_1 is to form according to 3.30 with ℓ from 3.31 (τ as in 3.32). Moreover, between these ℓ and b there exists the relation

3.39 $$b = \frac{\ell - [\nu(z_0) + \mu(z_0)]}{1 - \ell\overline{[\nu(z_0) + \mu(z_0)]}}.$$

II.4 COMPACTNESS

With the aid of the representation theorem certain compactness criteria for analytic functions can be carried over to solutions of elliptic systems. Thus we have the following.

4.1 Theorem. *Let ν_n, μ_n be functions defined and measurable in $\mathbb{C}$ and satisfying*

$$|\nu_n(z)| + |\mu_n(z)| \leqq \textit{constant} =: k < 1 \quad \text{a.e. in } \mathbb{C} \quad \forall n = 1, 2, \ldots$$

Further, let $w_n(z) = f_n \circ \chi_n(z)$ be (ν_n, μ_n)-solutions in the (fixed) domain $G \subseteqq \mathbb{C}$, $n = 1, 2, \ldots$, where χ_n means a K-quasiconformal mapping of $\mathbb{C}$ onto itself with $\chi_n(0) = 0$, $\chi_n(1) = 1$, $K = (1 + k)/(1 - k)$, and f_n is analytic in $\chi_n(G)$. The $w_n(z)$ are to be locally uniformly bounded in G, i.e.

$$|w_n(z)| \leqq K(G') \quad \forall n = 1, 2, \ldots, \quad \forall z \in G' \Subset G,$$

where the constant $K(G')$ depends only upon $G' \Subset G$.

(I) *Then the set of $w_n(z)$ is Hölder continuous with exponent $1 - (2/p)$, for every $p > 2$ satisfying $kC(p) < 1$, in each $G' \Subset G$.*

(II) *A subsequence of the $w_n(z)$ converges locally uniformly in G to a function $w(z) = f \circ \chi(z)$, where χ is a K-quasiconformal mapping of $\mathbb{C}$ onto itself with $\chi(0) = 0$, $\chi(1) = 1$, and f is analytic in $\chi(G)$.*

(III) *Every limit function $w(z)$ mentioned under* (II) *belongs to $W^1_{p,\text{loc}}(G)$ for every p as in* (I).

(IV) *The norms $\|w_{nz}\|_{L_p(G')}$, $n = 1, 2, \ldots$, are uniformly bounded for every $G' \Subset G$ and every p as in* (I).

(V) *If $\nu_n \to \nu$, $\mu_n \to \mu$ pointwise a.e. in G as $n \to \infty$, then every limit function $w(z)$ is a (ν, μ)-solution in G.*

Proof. The assertions in (I) and (II) follow from theorem I.3.20(II) and the locally uniform Lipschitz continuity of locally uniformly bounded analytic functions. (III) follows from (II), I.2.29 and chain rule I (0.3.23). As to the proof of (IV), let G' be any domain $\Subset G$. We fix a second domain G''

such that $G' \Subset G'' \Subset G$ and put

4.2 $\sigma_n(z) = w_{n\bar{z}}/w_{nz}$ a.e. in G'', $\sigma_n(z) = 0$ otherwise in $\mathbb{C}$

(cf. theorem 2.17). Then we have

4.3 $$w_n(z) = F_n \circ f^{h,\sigma_n}(z) \quad \text{in } G'',$$

where F_n is analytic in $f^{h,\sigma_n}(G'')$. Because of the uniform boundedness of the $w_n(z)$ in G'' and the compactness of the f^{h,σ_n} (cf. I.3.20(I)), the derivatives of the F_n are uniformly bounded in $f^{h,\sigma_n}(G')$, which together with I.2.9 implies the assertion (IV).

To prove (V) let A be any measurable set $\Subset G$. After changing over to a subsequence, if necessary, we may assume that w_n converges locally uniformly in G to $w(z)$. Clearly, we have

4.4 $$\begin{aligned} 0 &= \lim_{n\to\infty} \iint_A (w_{n\bar{z}} - \nu_n w_{nz} - \mu_n \overline{w_{nz}})\, d\sigma \\ &= \lim_{n\to\infty} \left(\iint_A (w_{n\bar{z}} - \nu w_{nz} - \mu \overline{w_{nz}})\, d\sigma + \iint_A (w_{nz}(\nu - \nu_n) + \overline{w_{nz}}(\mu - \mu_n))\, d\sigma \right). \end{aligned}$$

By 0.3.5 in conjunction with (IV) proved just now we obtain

$$\lim_{n\to\infty} \iint_A (w_{n\bar{z}} - \nu w_{nz} - \mu \overline{w_{nz}})\, d\sigma = \iint_A (w_{\bar{z}} - \nu w_z - \mu \overline{w_z})\, d\sigma .$$

Further, the Hölder inequality, (IV), and the theorem of Lebesgue on bounded convergence yield

$$\lim_{n\to\infty} \iint_A (w_{nz}(\nu - \nu_n) + \overline{w_{nz}}(\mu - \mu_n))\, d\sigma = 0 .$$

Since A is arbitrary we infer $w_{\bar{z}} - \nu w_z - \mu \overline{w_z} = 0$ a.e. in G, and because of (III) $w(z)$ is especially a (L_2-) solution of the (ν, μ)-system in G. Theorem 4.1 is proved.

II.5 DIFFERENTIABILITY, A-PRIORI-ESTIMATES AT INTERIOR POINTS

Next we prove some differentiability properties of (ν, μ)-solutions under certain additional smoothness conditions on the coefficients ν, μ (cf. [B4], p. 499, [L/V], pp. 233–236).

5.1 ***Definition.*** Let $G \subseteq \mathbb{C}$ be any measurable set and D an arbitrary subset of G. By $HL_p(D, G)$ we denote the set of all functions f defined and measurable in G for which

(I) $\dfrac{f(z) - f(z_0)}{z - z_0} \in L_p(G)$ (as a function of z) for each $z_0 \in D$

and

(II) $$\|f\|_{HL_p(D,G)} := \|f\|_{L_\infty(G)} + \sup\left\{ \left\| \frac{f(z) - f(z_0)}{z - z_0} \right\|_{L_p(G)} : z_0 \in D \right\} < \infty .$$

If $G = \mathbb{C}$ we only write $HL_p(D)$ instead of $HL_p(D, \mathbb{C})$.

For instance, if G is a domain $\subseteqq \mathbb{C}$ and $C_0^{0,\alpha}(G)$ denotes the space of Hölder continuous functions with exponent α, $0 < \alpha < 1$, and with support compactly contained in G, then obviously $C_0^{0,\alpha}(G) \subset HL_p(G, G)$ for every p such that $2 < p < 2/(1 - \alpha)$.

5.2 Theorem. (I) *Let $\mathfrak{P}$ be a set of pairs (ν, μ). For every $(\nu, \mu) \in \mathfrak{P}$ let ν, μ be functions from $HL_p(D, G)$, where G is a domain, $p > 2$. These ν, μ are to be defined and measurable everywhere in $\mathbb{C}$ and*

(1) $|\nu(z)| + |\mu(z)| \leqq$ *constant* $=: k < 1$ $\forall z \in \mathbb{C}$, $\|\nu\|_{HL_p(D,G)} + \|\mu\|_{HL_p(D,G)} \leqq K$ *(with the same constants k, K, p and the same sets D, G for every $(\nu, \mu) \in \mathfrak{P}$). Furthermore, let $\mathfrak{M}$ be any set of functions $w(z)$ such that every $w \in \mathfrak{M}$ is a (ν, μ)-solution for at least one $(\nu, \mu) \in \mathfrak{P}$, and for every set $G' \Subset G$ there exists a constant $K(G')$ such that*

(2) $|w(z)| \leqq K(G')$ $\forall z \in G'$ *and* $\forall w \in \mathfrak{M}$.

(II) *Then $\mathfrak{M}$ is equidifferentiable at every point $z_0 \in D$ and uniformly equidifferentiable in every set D' satisfying $D' \subseteqq D$, $D' \Subset G$. Every (ν, μ)-solution $w \in \mathfrak{M}$ with $(\nu, \mu) \in \mathfrak{P}$ satisfies its (ν, μ)-system not only a.e. in D, but everywhere in D.*

(III) *The derivatives w_z, $w_{\bar{z}}$ of the $w \in \mathfrak{M}$ are uniformly bounded in D', and their restrictions $w_z|_{D'}$, $w_{\bar{z}}|_{D'}$ to D' constitute a subset of $\mathscr{C}(D')$ which is equicontinuous in D', where D' is any set as in* (II).

(IV) *The derivatives w_z, $w_{\bar{z}}$ of the $w \in \mathfrak{M}$ constitute a bounded set in $HL_{p^*}(D', D')$ for every measurable D' with $D' \subseteqq D$, $D' \Subset G$ and for certain $p^* > 2$.*

(V) *If additionally $\nu, \mu \in \mathscr{C}(D, G)$ for every $(\nu, \mu) \in \mathfrak{P}$ and if the set of these ν, μ is equicontinuous in every D' as in* (IV), *then* (IV) *holds with every $p^* \in (2, p)$.*

(VI) *For every $(\nu, \mu) \in \mathfrak{P}$ let ν, μ satisfy the Hölder condition*

(3) $|\nu(z) - \nu(z_0)| \leqq K^*|z - z_0|^\alpha$, $|\mu(z) - \mu(z_0)| \leqq K^*|z - z_0|^\alpha$ $\forall z \in \mathbb{C}$, $\forall z_0 \in D$, *where the constants K^* and $\alpha \in (0, 1)$ are to be independent of the pair (ν, μ) at the time. Then for every $w \in \mathfrak{M}$ and every D' as in* (II) *it holds* $|w_z(z_1) - w_z(z_2)| \leqq K(\alpha', D')\,|z_1 - z_2|^{\alpha'}$, $|w_{\bar{z}}(z_1) - w_{\bar{z}}(z_2)| \leqq K(\alpha', D')\,|z_1 - z_2|^{\alpha'}$ $\forall z_1, z_2 \in D'$, *where the constant α' is arbitrary subject to $\alpha' \in (0, \alpha)$ and the constant $K(\alpha', D')$ depends on α' and D', but neither on $(\nu, \mu) \in \mathfrak{P}$ nor on $w \in \mathfrak{M}$.*

Proof. Let D' be any set mentioned in (II), G' be any open set such that $D' \Subset G' \Subset G$, and d be any positive number satisfying $\operatorname{dist}(D', \partial G') \geqq 2d$, $d \leqq 1$. Further, we fix any $(\nu, \mu) \in \mathfrak{P}$, any corresponding (ν, μ)-solution $w \in \mathfrak{M}$ and any point $z_0 \in D'$.

With b, $\mathscr{b}$ to be any constants of absolute value less than one, we put

5.3 $$g(z) := g(z; z_0) := \frac{w(z) - w(z_0) + b\overline{(w(z) - w(z_0))}}{z - z_0 + \mathscr{b}\overline{(z - z_0)}}.$$

This g is a solution of

5.4 $$g_{\bar z} = \nu_0 g_z + \mu_0 \overline{g_z} + A_0 g + B_0 \bar g \quad \text{in } \{0 < |z - z_0| < 2d\}$$

with

5.5 $$\nu_0 = \nu \cdot (1 - |b|^2)/\mathfrak{N}_0, \quad \mathfrak{N}_0 = |1 + \bar\mu b|^2 - |\nu b|^2, \quad \nu = \nu(z),\ \mu = \mu(z),$$

$$\mu_0 = \frac{\overline{z - z_0} + \bar{\ell}(z - z_0)}{z - z_0 + \ell(\overline{z - z_0})} (\bar\mu b^2 + (1 + |\mu|^2 - |\nu|^2)\, b + \mu)/\mathfrak{N}_0,$$

$$A_0 = (\nu_0 - \ell)/(z - z_0 + \ell(\overline{z - z_0})), \qquad B_0 = \mu_0/(\overline{z - z_0} + \bar\ell(z - z_0))$$

for $|z - z_0| < 2d$. In order to define the coefficients of the system 5.4 everywhere in $\mathbb{C}$ we (arbitrarily) put

5.6 $$\nu_0(z) = \mu_0(z) = A_0(z) = B_0(z) = 0 \quad \text{for } |z - z_0| \geqq 2d.$$

Now we choose

5.7 $$b = b(z_0) = 2\varkappa/(1 + \sqrt{1 - 4|\varkappa|^2}) \text{ with } \varkappa = -\mu(z_0)/(1 + |\mu(z_0)|^2 - |\nu(z_0)|^2),$$
$$\ell = \ell(z_0) = \nu_0(z_0).$$

Then $A_0(z)$, $B_0(z) \in L_p$ because of $z_0 \in D' \subseteqq D$ and 5.2(1). Furthermore we have

5.8 $$|b(z)| \leqq k, \ |\ell(z)| \leqq k, \ |\nu_0(z)| + |\mu_0(z)| \leqq \text{constant} =: k' < 1 \quad \forall z \in \mathbb{C}$$

(cf. section II.3, in particular 3.32, 3.36) and

5.9 $$\|A_0(z)\|_{L_p} + \|B_0(z)\|_{L_p} \leqq \text{constant} =: K' \quad \text{for each } z_0 \in D',$$

where k' depends only on k, and K' only on k and K. The function g, with ℓ and b from 5.7, admits the representation

5.10 $$g(z; z_0) = e^{s(z)} f \circ \chi(z) \quad \text{in } \{0 < |z - z_0| < 2d\},$$

by the representation theorem. Here $s(z) = s(z; z_0)$ is a function of z in $\mathbb{C}$ satisfying

5.11 $$|s(z)| \leqq S_1, \qquad |s(z_1) - s(z_2)| \leqq S_2 |z_1 - z_2|^{1-(2/p')} \quad \forall z, z_1, z_2 \in \mathbb{C},$$

where p' is any fixed number subject to $k'C(p') < 1$ and $2 < p' \leqq p$. The constants S_1, S_2 are independent of $z_0 \in D'$, of the choice of D' (note $d \leqq 1$) and of $w \in \mathfrak{M}$. They depend only on p', k' and K' (cf. 2.12, 2.14, 5.5, 5.6) and hence, in the long run, only on p', k and K. The function $\chi(z) = \chi(z; z_0)$ in 5.10 is a hydrodynamically normalized schlicht Beltrami solution in $\mathbb{C}$ satisfying

5.12 $$|\chi(z_1) - \chi(z_2)| \leqq S_3 |z_1 - z_2|^{1-(2/p')} \quad \forall z_1, z_2 \in \{|z| < R\},$$
$$|z(\chi_1) - z(\chi_2)| \leqq S_4 |\chi_1 - \chi_2|^{1-(2/p')} \quad \forall \chi_1, \chi_2 \in \{|\chi| < 2R\},$$

where $z(\chi)$ is the inverse of $\chi(z)$ and R is any positive constant such that $G' \subseteqq \{|z| < R\}$. Here the constants S_3, S_4 may be chosen to be $\geqq 1$, while otherwise depending only on p', k' and R and hence, in the end, only on p', k and $\sup \{|z| : z \in G'\}$ (cf. 5.5, 2.20, I.2.10, I.2.21).

The function $f(\chi) = f(\chi; z_0)$ is known to be analytic first only for $\chi \in \chi(\{0 < |z - z_0| < 2d\}) = \chi(\cdot; z_0)(\{0 < |z - z_0| < 2d\})$. Since

5.13 $$|f(\chi)| \leqq e^{S_1} \frac{1+k}{1-k} \left| \frac{w(z(\chi)) - w(z(\chi_0))}{z(\chi) - z(\chi_0)} \right| \leqq \text{constant} \cdot |\chi - \chi_0|^{-p'/(p'-2)}$$

in a neighbourhood of $\chi_0 := \chi(z_0; z_0)$ (cf., for example, 4.1 (I)), χ_0 is not an essential singularity. Neither is it a pole, because of the argument principle and

5.14 $$\Delta_{\mathfrak{C}} \log f(\chi) = \Delta_{\mathfrak{C}} \log \frac{w(z) - w(z_0)}{z - z_0} \geqq 0, \quad z = z(\chi),$$

where $\mathfrak{C}$ is a positively oriented circle $\{|\chi - \chi_0| = r\}$ with r sufficiently small so that $f(\chi)$ does not possess any zero in $\{0 < |\chi - \chi_0| \leqq r\}$. Hence, χ_0 is always a removable singularity of f. This means that g can be extended to a solution of 5.4 in $\{|z - z_0| < 2d\}$.

By 5.12 we have

5.15 $$\{|\chi - \chi_0| < 2d'\} \subseteqq \chi(\{|z - z_0| < d\}), \ \chi(\{|z - z_0| < d_0\}) \subseteqq \{|\chi - \chi_0| < d'\}$$

with $2d' := (d/S_4)^{p'/(p'-2)}$, $d_0 := (d'/S_3)^{p'/(p'-2)}$.

By the maximum principle for analytic functions, $|f(\chi)|$ assumes its maximum in $\chi(\{|z - z_0| \leqq d\})$ at the boundary $\chi(\{|z - z_0| = d\})$. Together with 5.13 this gives

5.16 $$|f(\chi(z))| \leqq e^{S_1} \frac{1+k}{1-k} \frac{2K(G')}{d} =: K_1 \quad \text{as } |z - z_0| \leqq d,$$

hence

5.17 $$|f(\chi)| \leqq K_1 \quad \text{for } |\chi - \chi_0| < 2d', \qquad |f'(\chi)| \leqq \frac{K_1}{d'} \quad \text{for } |\chi - \chi_0| < d',$$

by 5.15 and the Cauchy integral formula (for $f'(\chi)$).

(By the way, if $w(z)$ is schlicht in $\{|z - z_0| < 2d\}$, then equality occurs in 5.14 and, consequently, $f(\chi)$ is analytic and different from zero in $\chi(\{|z - z_0| < 2d\})$. Again by the maximum principle for analytic functions, applied to $1/f(\chi)$, we obtain

5.18 $$|f(\chi(z))| \geqq e^{-S_1} \frac{1-k}{1+k} d^{-1} \min \{|w(z) - w(z_0)| : |z - z_0| = d\} =: K_2^* \quad \text{for } |z - z_0| \leqq d;$$

we shall come back to 5.18 in the proof of theorem 5.47.)

By 5.17 it then holds

5.19 $$\begin{aligned} |g(z) - g(z_0)| &\leqq |e^{s(z)} - e^{s(z_0)}| \, |f \circ \chi(z)| + |e^{s(z_0)}| \, |f \circ \chi(z) - f \circ \chi(z_0)| \\ &\leqq e^{S_1} \cdot S_2 K_1 |z - z_0|^{1-(2/p')} + e^{S_1} \cdot S_3 (K_1/d') \, |z - z_0|^{1-(2/p')} \end{aligned}$$

for $|z - z_0| < d_0$. Because of 5.2(2), 5.3, from this we further obtain

5.20 $$|g(z; z_0) - g(z_0; z_0)| \leqq K_3 |z - z_0|^{1-(2/p')} \quad \text{for every } z_0 \in D' \text{ and every } z \in G'$$

with a constant K_3 depending only on p', k, K, G' and d (concerning G', K_3 depends on $K(G')$ and $\sup\{|z| : z \in G'\}$). Thus we have

5.21 $$w(z) - w(z_0) + b(\overline{w(z) - w(z_0)}) = g(z_0)\,(z - z_0 + \ell(\overline{z - z_0})) + O(|z - z_0|^{2-(2/p')}),$$

hence

5.22 $$w(z) = w(z_0) + \frac{g(z_0) - b\ell\overline{g(z_0)}}{1 - |b|^2}(z - z_0) + \frac{g(z_0)\,\ell - b\overline{g(z_0)}}{1 - |b|^2}(\overline{z - z_0}) + O(|z - z_0|^{2-(2/p')})$$

with $b = b(z_0)$, $\ell = \ell(z_0)$ and

5.23 $$|O(|z - z_0|^{2-(2/p')})| \leqq \frac{1 + k}{1 - k} K_3|z - z_0|^{2-(2/p')} =: K_4|z - z_0|^{2-(2/p')}$$

for every $z_0 \in D'$, every $z \in G'$ and every $w \in \mathfrak{M}$. The coefficients of $z - z_0$ and $\overline{z - z_0}$ are, of course, $w_z(z_0)$ and $w_{\bar z}(z_0)$, respectively. After a simple calculation, whilst observing 5.7, this provides the relation $w_{\bar z}(z_0) = \nu(z_0)\,w_z(z_0) + \mu(z_0)\,\overline{w_z(z_0)}$ for every $z_0 \in D$. Assertion 5.2(II) is proved.

From 5.22, 5.23 we further conclude

5.24 $$|w_z(z_0)| \cdot (1 - k) = \left|\frac{w(z) - w(z_0)}{z - z_0}\right| + K_4|z - z_0|^{1-(2/p')}$$

for every $z_0 \in D'$, every $z \in G'$ and every $w \in \mathfrak{M}$. Together with 5.2(2) and $d \leqq 1$ this yields

5.25 $$|w_{\bar z}(z_0)| \leqq |w_z(z_0)| \leqq \frac{2K(G') + K_4}{d(1 - k)} \quad \forall z_0 \in D'.$$

This proves that w_z, $w_{\bar z}$ are uniformly bounded in D' for all $w \in \mathfrak{M}$, which is the first assertion of 5.2(III).

Now, let z_1, z_2 be any two points of D'. For any $z_1', z_2' \in G'$ we then have

5.26 $$|w(z_j') - w(z_j) - (z_j' - z_j)\,w_z(z_j) - (\overline{z_j' - z_j})\,w_{\bar z}(z_j)| \leqq K_4|z_j' - z_j|^{2-(2/p')}, \quad j = 1, 2.$$

We now choose z_1', z_2' such that $z_1' - z_1 = z_2' - z_2 = \text{real} =: h \in (0, d]$. On summing up both inequalities 5.26 (for $j = 1$ and $j = 2$) and observing $w_z + w_{\bar z} = w_x$ we obtain

5.27 $$|w_x(z_1) - w_x(z_2)| \leqq 2K_4\,h^{1-(2/p')} + \frac{|w(z_1) - w(z_2)|}{h} + \frac{|w(z_1') - w(z_2')|}{h}.$$

This implies the asserted equicontinuity in D' of the set of restrictions $w_x|_{D'}$, $w \in \mathfrak{M}$. For let us fix any positive ε. Then we can choose h (>0 and $\leqq d$) so that $2K_4h^{1-(2/p')} < \varepsilon/2$. Because of 5.2(2) and theorem 4.1(I) there is a positive $\delta(\varepsilon)$ such that both the remaining terms on the right-hand side are also less than $\varepsilon/2$ for every $w \in \mathfrak{M}$ if $|z_1 - z_2| = |z_1' - z_2'| < \delta(\varepsilon)$ and as long as z_1, z_2, z_1', z_2' belong to G'; the latter is satisfied at any rate if $z_1, z_2 \in D'$ and $h \leqq d$. Analogously the equicontinuity of the set of $w_y|_{D'}$ in D' follows. Assertion 5.2(III) is proved. (Of course, the statement concerning equicontinuity loses its purpose when D consists only of isolated points).

In proving 5.2(IV) we put $z_1' = z_2$, $z_2' = z_1$ in 5.26. When summing up both the inequalities 5.26 and applying the triangle inequality we obtain

5.28 $$\big||w_z(z_1) - w_z(z_2)| - |w_{\bar z}(z_1) - w_{\bar z}(z_2)|\big| \leqq 2K_4|z_1 - z_2|^{1-(2/p')}$$

for all $z_1, z_2 \in D'$ and every $w \in \mathfrak{M}$. By the last statement in 5.2(II) w fulfils each associated (ν, μ)-system with $(\nu, \mu) \in \mathfrak{P}$ at z_1 and z_2. From this and 5.28 it follows

5.29 $$(1 - k)\,|w_z(z_1) - w_z(z_2)| \leqq 2K_4|z_1 - z_2|^{1-(2/p')} + |w_z(z_1)|\,(|\nu(z_1) - \nu(z_2)| + |\mu(z_1) - \mu(z_2)|) \quad \forall z_1, z_2 \in D',\ w \in \mathfrak{M}.$$

This, together with 5.25 and (concerning $w_{\bar z}$) 5.28 once more, yields the assertion 5.2(IV) for every p^* satisfying $2 < p^* < p'$.

We now come to the proof of 5.2(V). Let $T(z_0, \ell)$ be the affine mapping of the z-plane

5.30 $$t = z + \ell(\overline{z - z_0}), \quad |\ell| < 1,$$

and $\mathscr{T}(R, k)$ be the set of all the mappings $T(z_0, \ell)$ with $|z_0| \leqq R$, $|\ell| \leqq k < 1$ (with the same k as in 5.2(1)). As above, we again fix G', D' such that $D' \Subset G' \Subset G$, $D' \subseteqq D$. The mapping $T(z_0, \ell)$ carries D', G', G into D_t', G_t', G_t, respectively, and we have

5.31 $$\operatorname{dist}(\partial G_t', D_t') \geqq \text{constant} =: 2\delta^* > 0 \quad \forall T(z_0, \ell) \in \mathscr{T}(R, k).$$

We now fix any $w \in \mathfrak{M}$, an associated $(\nu, \mu) \in \mathfrak{P}$ and any $z_0 \in D'$. The mapping $T(z_0, \ell)$ carries $w(z)$ into

5.32 $$f(t) := w(z(t)), \quad z(t) = z_0 + \frac{1}{1 - |\ell|^2}(t - z_0) - \frac{\ell}{1 - |\ell|^2}(\overline{t - z_0}).$$

This $f(t)$ is a solution of

5.33 $$f_{\bar t} = \nu_1(t)\, f_t + \mu_1(t)\, \overline{f_t} \quad \text{in } G_t,$$

where

5.34 $$\nu_1(t) = \frac{\bar\nu \ell^2 - (1 + |\nu|^2 - |\mu|^2)\,\ell + \nu}{\mathfrak{N}_1},$$

$$\mu_1(t) = \mu\frac{1 - |\ell|^2}{\mathfrak{N}_1}, \quad \mathfrak{N}_1 = |1 - \bar\nu\ell|^2 - |\mu\ell|^2, \quad \nu = \nu(z(t)), \quad \mu = \mu(z(t)).$$

We now choose

5.35 $$\ell = \ell(z_0) = \frac{2\sigma}{1 + \sqrt{1 - 4|\sigma|^2}}, \qquad \sigma = \frac{\nu(z_0)}{1 + |\nu(z_0)|^2 - |\mu(z_0)|^2}.$$

For every $(\nu, \mu) \in \mathfrak{P}$ and every z_0 $(\in \mathbb{C})$ it then holds

5.36 $$|\ell| \leqq k,$$

cf. 3.32. With this ℓ in 5.34 we have $\nu_1(z_0) = 0$. With the corresponding $f(t)$ we now put

5.37 $$g(t) = g(t; z_0) = \frac{f(t) - f(z_0) + b(\overline{f(t) - f(z_0)})}{t - z_0}$$

(where z_0 is the same point as in 5.30). Let $\nu_0(t)$, $\mu_0(t)$, $A_0(t)$, $B_0(t)$ be those functions which we obtain according to 5.5 by replacing there the $\nu(z)$, $\mu(z)$ by $\nu_1(t)$, $\mu_1(t)$, respectively (note that the ℓ, which occurs in 5.5, 5.7, is now equal to zero). As to b in 5.37, in keeping with 5.7 we put

5.38 $$b = b(z_0) = \frac{2\varkappa}{1 + \sqrt{1 - 4|\varkappa|^2}}, \qquad \varkappa = \frac{-\mu_1(z_0)}{1 + |\mu_1(z_0)|^2},$$

i.e. $b = -\mu_1(z_0)$. Then we have $\nu_0(z_0) = \mu_0(z_0) = 0$ and also $|b(z_0)| \leqq k < 1$, cf. 3.36 and the subsequent remarks until the end of section II.3.

We now choose $\varepsilon > 0$ such that

5.39 $$\varepsilon C(p) < 1, \quad p \text{ from } 5.2(1).$$

By hypothesis, $\nu, \mu \in \mathscr{C}(D, G)$ for every $(\nu, \mu) \in \mathfrak{P}$, and these ν, μ are to form a set which is equicontinuous in each D'. Hence, there is a $\delta > 0$ such that

5.40 $$|\nu_0(t)| + |\mu_0(t)| \leqq \varepsilon \quad \text{if } |t - z_0| < 2\delta$$

for every $(\nu, \mu) \in \mathfrak{P}$, every $T(z_0, \ell(z_0)) \in \mathscr{T}(R, k)$ and every $z_0 \in D'$. In addition, we may require $\delta \leqq \delta^*$, cf. 5.31, and we now put

$$\nu_0(t) = \mu_0(t) = A_0(t) = B_0(t) = 0 \quad \text{as } |t - z_0| \geqq 2\delta.$$

For the coefficients $A_0(t)$, $B_0(t)$ we then again have an estimate

5.41 $$\|A_0\|_{L_p} + \|B_0\|_{L_p} \leqq K \cdot K_0 =: K'$$

with one and the same constant K_0 for every $z_0 \in D'$, every $(\nu, \mu) \in \mathfrak{P}$ and every $T(z_0, \ell(z_0))$.

For $g(t; z_0)$ (with b from 5.38) we now again have a representation

5.42 $$g(t; z_0) = e^{s(t)} f \circ \chi(t),$$

where s, f, χ possess all the properties mentioned in 5.11 to 5.18, but where we now have $p' = p$ (because $k' = \varepsilon$) and where d has to be replaced by δ. We again obtain an estimate

5.43 $$|g(t; z_0) - g(z_0; z_0)| \leqq K_3 |t - z_0|^{1-(2/p)} \quad \forall z_0 \in D', \ \forall t \in G'_t,$$

at present with the original p from 5.2(1), instead of p'. Here the constant K_3 depends only on p, k, K, G' and δ. From this we further infer

$$f(t) = f(z_0) + \frac{g(z_0; z_0)}{1 - |b(z_0)|^2}(t - z_0) - \frac{b(z_0)\,\overline{g(z_0; z_0)}}{1 - |b(z_0)|^2}\overline{(t - z_0)}$$
$$+ O(|t - z_0|^{2-(2/p)}),$$

and this implies a relation of the form

5.44 $$w(z) = w(z_0) + a_1(z_0)(z - z_0) + a_2(z_0)(\overline{z - z_0}) + O(|z - z_0|^{2-(2/p)})$$

with

5.45 $$|O(|z - z_0|^{2-(2/p)})| \leqq K_4|z - z_0|^{2-(2/p)} \quad \forall z_0 \in D', \forall z \in G', \forall w \in \mathfrak{M},$$

where the constant K_4 depends only on p, k, K, G' and δ (and δ depends only on k, on the 'module of continuity' of the ν, μ, i.e. the relation between ε and δ in 5.40, and on dist $(\partial G', D')$). From 5.44, 5.45, as in the proof of 5.2(IV), the relation 5.29 follows as does a corresponding relation for $w_{\bar{z}}$ with $p' = p$, which proves 5.2(V).

If condition 5.2(3) is satisfied, in 5.2(1) we may take $p = 2/(1 - \alpha')$ for any $\alpha' \in (0, \alpha)$. Then 5.29 reads

5.46 $$(1 - k)|w_z(z_1) - w_z(z_2)| \leqq 2K_4|z_1 - z_2|^{\alpha'} + |w_z(z_1)|\, 2K^*|z_1 - z_2|^{\alpha} \quad \forall z_1, z_2 \in D', \forall w \in \mathfrak{M},$$

and 5.46, together with 5.28, yields a corresponding inequality for $w_{\bar{z}}$, which, in conjunction with 5.2(III), proves 5.2(VI).

As a completion to theorem 5.2 we have

5.47 Theorem. *Let $\mathfrak{P}$ and $\mathfrak{M}$ satisfy the assumptions specified in* 5.2(I), *let every $w \in \mathfrak{M}$ be schlicht in G and $\mathfrak{M}$ be relatively compact in the set of $(1 + k)/(1 - k)$-quasiconformal mappings of G (with respect to locally uniform convergence in G). Then for each D' mentioned in* 5.2(II) *there exists a constant $c(D') > 0$ such that*

(I) $|w_z(z_0)|^2 - |w_{\bar{z}}(z_0)|^2 \geqq c(D') \; \forall z_0 \in D', \forall w \in \mathfrak{M}$.

Proof. Since we have $w_{\bar{z}}(z_0) = \nu(z_0)\, w_z(z_0) + \mu(z_0)\, \overline{w_z(z_0)}$ for every $z_0 \in D$, it is enough to prove the existence of a positive constant $q(D')$ such that $|w_z(z_0)| \geqq q(D') \; \forall z_0 \in D'$ and $\forall w \in \mathfrak{M}$.

Again fix an open set G' satisfying $D' \Subset G' \Subset G$, let z_0 be any fixed point from D' and d be a positive constant as in the proof of 5.2. Because $\mathfrak{M}$ is relatively compact within the set of $(1 + k)/(1 - k)$-quasiconformal mappings of G we have

$$d_1 := \inf \{\min \{|w(z) - w(z_0)| : |z - z_0| = d\} : w \in \mathfrak{M}, z_0 \in D'\} > 0,$$

hence, by 5.18,

$$|f \circ \chi(z_0)| \geqq e^{-S_1} \frac{1 - k}{1 + k} \frac{d_1}{d} =: K_2 > 0, \qquad |g(z_0; z_0)| \geqq K_2\, e^{-S_1}.$$

According to 5.22 we have

$$w_z(z_0) = (1 - |b|^2)^{-1} (g(z_0; z_0) - b\overline{\ell g(z_0; z_0)}),$$

therefore

5.48 $$|w_z(z_0)| \geqq (1 - k^2) K_2\, e^{-S_1} =: q(D') > 0 \quad \forall z_0 \in D', \forall w \in \mathfrak{M},$$

which we had to prove.

The theorems 5.2 and 5.47 retrospectively yield a statement on $HL_p(D, G)$, namely

5.49 Theorem. *For every $h \in HL_p(D, G)$ with $p > 2$ the restriction $h|_D$ is continuous in D. The imbedding*

$$\varphi : HL_p(D, G) \to \mathscr{C}(D'), \quad \varphi h := h|_{D'}$$

is compact for every D' subject to $D' \subseteq D$, $D' \Subset G$.

Proof. Without loss of generality let $\|h\|_{L_\infty(G)} < 1$. We put $\nu(z) = h(z)$ if $z \in G$, $\nu(z) = 0$ otherwise. By 5.2(III) $f_z^\nu, f_{\bar z}^\nu$ (cf. I.2.1) are continuous in D and, by 5.47, $f_z^\nu \neq 0$ in D. Consequently, $\nu(z) = f_{\bar z}^\nu / f_z^\nu$ is continuous in D.

Next, let $\mathscr{H}$ be a bounded set $\subset HL_p(D, G)$, and without loss of generality let $\|h\|_{L_\infty(G)} \leqq \text{constant} =: k < 1 \ \forall h \in \mathscr{H}$. We choose $\mathfrak{P}$ to be the set of pairs $(\nu, 0)$ with $\nu|_G \in H$, $\nu = 0$ outside G, and $\mathfrak{M} = \{f^\nu : (\nu, 0) \in P\}$. By theorem I.3.20(II) $\mathfrak{M}$ satisfies the conditions of theorem 5.47. By 5.2(III) the $f_z^\nu, f_{\bar z}^\nu$ are uniformly bounded and equicontinuous in D' and, by 5.47, $|f_z^\nu| \geqq \text{constant} > 0$ in each D'. Hence, the set of $\nu(z) = f_{\bar z}^\nu / f_z^\nu$ is also equicontinuous (and, of course, bounded) in D'. By the Ascoli–Arzelá theorem (cf. [Y], p. 85), this implies the last assertion in 5.49. Theorem 5.49 is proved.

II.6 ADJOINT SYSTEMS, THE CAUCHY INTEGRAL THEOREM

Let $f(z)$ be any (ν, μ)-solution in the domain $G \subseteq \mathbb{C}$, and D be any subdomain of G as in 0.3.17. Let the function $h(z)$ satisfy the conditions of 0.3.17. Then we have (cf. 0.3.19)

6.1

$$\operatorname{Re} \frac{1}{2i} \oint_{\partial D} f \, dh = -\operatorname{Re} \iint_D (f_z h_{\bar z} - f_{\bar z} h_z) \, d\sigma = -\operatorname{Re} \iint_D f_z (h_{\bar z} - \nu h_z - \bar\mu \overline{h_z}) \, d\sigma .$$

In the same way we have

6.2

$$\operatorname{Im} \frac{1}{2i} \oint_{\partial D} f \, dh = -\operatorname{Im} \iint_D (f_z h_{\bar z} - f_{\bar z} h_z) \, d\sigma = -\operatorname{Im} \iint_D f_z (h_{\bar z} - \nu h_z + \bar\mu \overline{h_z}) \, d\sigma .$$

We call the system

6.3 $$h_{\bar z} = \nu h_z + \bar\mu \overline{h_z}$$

the adjoint system and the system

6.4 $$h_{\bar z} = \nu h_z - \bar\mu \overline{h_z}$$

the $(-)$-adjoint system of $f_{\bar z} = \nu f_z + \mu \overline{f_z}$. Then the following generalization of the Cauchy integral theorem holds.

6.5 Theorem (generalized Cauchy integral theorem, cf. [S1], [P2], p. 56). *Let G be a domain $\subseteq \mathbb{C}$, D be a subdomain of G as in 0.3.17, f be a solution* *of the (ν, μ)-system in G, h be a solution of the adjoint $(\nu, \bar{\mu})$-system in G, and $\nu, \mu \in HL_p(\partial D, G)$, $p > 2$. Then it holds*

$$\operatorname{Re} \frac{1}{2i} \int_{\partial D} f \, dh = 0.$$

Proof. By 5.2(II), (III) h is absolutely continuous on ∂D, hence 6.1 holds, and this implies the assertion in 6.5.

A counterpart of 6.5 for $\operatorname{Im} \frac{1}{2i} \int_{\partial D} f \, dh$ follows by observing that ih is a $(\nu, \bar{\mu})$-solution if h is a $(\nu, -\bar{\mu})$-solution.

The smoothness condition $\nu, \mu \in HL_p(\partial D, G)$ is only necessary in order to ensure that h is of bounded variation on ∂D. Hence, we also have the

6.6 Theorem. *Let G, D, f be given as in theorem 6.5, h be of bounded variation on ∂D, $h \in W^1_{2,\mathrm{loc}}(G)$, and let h be a $(\nu, \bar{\mu})$-solution in D. Then the assertion of theorem 6.5 holds.*

Of course, the question arises, how the adjointness and $(-)$-adjointness defined above is connected with the usual notion of adjointness. For this let $\nu, \mu \in \mathscr{C}^1(G)$. The (ν, μ)-system can also be written as

6.7
$$L \begin{pmatrix} u \\ v \end{pmatrix} \equiv \begin{pmatrix} v_x + cu_x + du_y \\ -v_y + au_x + bu_y \end{pmatrix} = \begin{pmatrix} 0 \\ 0 \end{pmatrix},$$

where a, b, c, d correspond to ν, μ according to 1.13 and belong also to $\mathscr{C}^1(G)$. With M to denote the adjoint differential operator to L, it must hold, by definition (cf. [C], p. 235),

6.8
$$(\varphi, \psi) L \begin{pmatrix} u \\ v \end{pmatrix} - (u, v) M \begin{pmatrix} \varphi \\ \psi \end{pmatrix} = \operatorname{div} P \quad \forall u, v, \varphi, \psi \in \mathscr{C}_0^\infty(G),$$

where $P = (p_1(x, y), p_2(x, y))$ has to be the null vector at $(x, y) \in G$ if at least one of the vectors (u, v), (φ, ψ) is the null vector at (x, y). The product rule leads to

6.9
$$M \begin{pmatrix} \varphi \\ \psi \end{pmatrix} = \begin{pmatrix} -(c\varphi + a\psi)_x - (d\varphi + b\psi)_y \\ -\varphi_x + \psi_y \end{pmatrix}.$$

Then

$$M \begin{pmatrix} \varphi \\ \psi \end{pmatrix} = \begin{pmatrix} 0 \\ 0 \end{pmatrix}$$

means in case of continuously differentiable φ, ψ that there are single-valued functions $X(x, y)$, $Y(x, y)$ satisfying $\varphi = X_y$, $\psi = X_x$ as well as

6.10
$$\begin{aligned} Y_x &= -bX_x - dX_y \\ Y_y &= aX_x + cX_y, \end{aligned}$$

and 6.10 is just the adjoint system of 6.7 in the sense of 6.3 (cf. 1.6).

The (−)-adjoint system will be obtained as follows. In 6.7 replace u by v and v by $-u$ (rotation by $-\pi/2$ of the coordinate system in the (u, v)-plane) and bring it again into the form

$$\tilde{L}\begin{pmatrix}u\\v\end{pmatrix} \equiv \begin{pmatrix}v_x + \tilde{c}u_x + \tilde{d}u_y\\-v_y + \tilde{a}u_x + \tilde{b}_y u\end{pmatrix} = \begin{pmatrix}0\\0\end{pmatrix}.$$

Let $\tilde{M}$ be the operator adjoint to $\tilde{L}$. Then the system

$$\tilde{M}\begin{pmatrix}\varphi\\\psi\end{pmatrix} = \begin{pmatrix}0\\0\end{pmatrix}$$

is just the system (−)-adjoint to 6.7.

The change-over from the (ν, μ)-system to its adjoint system means geometrically that the families of ellipses (α, β, γ), $(\alpha^*, \beta^*, \gamma^*)$ which correspond to the (ν, μ)-system according to 3.13, 3.14 change into (α, β, γ), $(\alpha^*, -\beta^*, \gamma^*)$, that is, the S^*-ellipses are replaced by their reflections at the real axis.

Chapter III

SOME FUNCTION THEORETIC PROPERTIES OF (ν, μ)-SOLUTIONS

By the representation theorem (ν, μ)-solutions (almost obviously) have many properties in common with analytic functions. The most important properties of this kind shall be presented in the first two sections of this chapter (of course, by the representation theorem many other theorems on analytic functions can easily be carried over to (ν, μ)-solutions). In the last two sections we deal with asymptotic behaviour of (ν, μ)-solutions at zeroes and poles.

III.1 ZEROES, ISOLATED SINGULARITIES

In line with the terminology used with analytic functions we will say:

1.1 *The function $w(z)$ is called a (ν, μ)-solution at the point $z_0 \in \mathbb{C}$ if $w(z)$ is a (ν, μ)-solution in a neighbourhood of z_0.*

1.2 ***Definition.*** Let $w(z)$ be a (ν, μ)-solution at z_0. The point z_0 will be called a zero of order n of $w(z)$ if in any representation

(I) $w(z) = f \circ \chi(z)$,

where χ is a quasiconformal mapping of a neighbourhood U of z_0 and f is analytic in $\chi(U)$, this f has a zero of order n at $\chi(z_0)$.

Of course, the order defined in this way of a zero of $w(z)$ does not depend on the representation 1.2(I) at the time. To see this let us consider two such representations

1.3 $$w(z) = f_1 \circ \chi_1(z) = f_2 \circ \chi_2(z) \quad \forall z \in U.$$

Since the zeroes of f_1 and f_2 are isolated in $\chi_1(U)$ and $\chi_2(U)$, respectively, there is a (positively oriented) circle $\mathfrak{C}$ with centre at z_0 and radius r such that $\chi_1(z_0)$ and $\chi_2(z_0)$ are the only zeroes of f_1 and f_2 in $\chi_1(\{|z - z_0| \leqq r\})$ and $\chi_2(\{|z - z_0| \leqq r\})$, respectively. Suppose $\chi_1(z_0)$ is a zero of order n of f_1, and $\chi_2(z_0)$ a zero of order m of f_2. For the increase of the argument of $w(z)$ along $\mathfrak{C}$ we then have

1.4 $$\Delta_{\mathfrak{C}} \arg w(z) = \Delta_{\chi_1(\mathfrak{C})} \arg f_1(\chi) = 2\pi n = \Delta_{\chi_2(\mathfrak{C})} \arg f_2(\chi) = 2\pi m,$$

hence $n = m$.

1.5 ***Definition.*** The point $z_0 \in \mathbb{C}$ is called an isolated singularity of the (ν, μ)-solution $w(z)$ if $w(z)$ is a (ν, μ)-solution in $\{0 < |z - z_0| < r\}$, where r is any positive number. Then $w(z)$ admits a representation 1.2(I), where χ is a quasiconformal mapping of $U = \{|z - z_0| < r\}$ and f is analytic in $\chi(U) \setminus \chi(z_0)$. The point z_0 is called a pole of order n, an essential singularity or a removable singularity of $w(z)$ if in any of these representations the analytic function f has a pole of order n, an essential singularity or a removable singularity, respectively, at $\chi(z_0)$.

On account of the behaviour of analytic functions in isolated singularities it is clear that neither the character of the isolated singularity z_0 of $w(z)$ nor the order of the pole (if z_0 is a pole) depend on the representation at the time. Also it is clear (by II.1.10 and 0.3.23) that $w(z)$ can be extended to a (ν, μ)-solution at z_0 (in the sense of 1.1) if z_0 is a removable singularity of $w(z)$.

In order that we no longer need to exclude the point ∞ we will agree:

1.6 *The point ∞ is called a removable singularity, a pole of order n or an essential singularity of the (ν, μ)-solution $w(z)$ if $w(z)$ is a (ν, μ)-solution in $\{R < |z| < \infty\}$ and admits a representation* 1.2(I) *where χ is a quasiconformal mapping of $\{|z| > R\}$ with $\chi(\infty) = \infty$ and f is an analytic function having a removable singularity, a pole of order n or an essential singularity, respectively, at ∞. If ∞ is a removable singularity of the (ν, μ)-solution $w(z)$, then we will also say that $w(z)$ is a (ν, μ)-solution at ∞.*

In an analogous sense it should be understood when we call ∞ a zero of order n of a (ν, μ)-solution $w(z)$.

1.7 *Let $G \subseteqq \overline{\mathbb{C}}$. We call $w(z)$ a (ν, μ)-solution in G if $w(z)$ is a (ν, μ)-solution at every point of G.*

(This terminology is in keeping with definition II.1.10, because $w(z)$ belongs to $W^1_{p,\mathrm{loc}}(G)$ if $w \in W^1_p(U_j)$ $\forall U_j$ with $\{U_j\}$ to be a countable set of open disks such that $U_j \Subset G$ and for each $G' \Subset G$ there is a finite number of U_j the union of which covers G', cf., e.g. [M3], theorem 3.1.2(b)).

As an immediate consequence of the behaviour of analytic functions in poles and removable singularities, in conjunction with the Casorati–Weierstraß theorem, we obtain the following.

1.8 Theorem. *Let z_0 be an isolated singularity of the (ν, μ)-solution $w(z)$.*

(I) *The point z_0 is a removable singularity of $w(z)$ if and only if*

$$\limsup_{z \to z_0} |w(z)| < \infty,$$

(II) *z_0 is a pole of $w(z)$ if and only if*

$$\lim_{z \to z_0} w(z) = \infty,$$

(III) *z_0 is an essential singularity of $w(z)$ if and only if for every $a \in \overline{\mathbb{C}}$ there is a sequence $z_{a,n} \to z_0$ as $n \to \infty$, so that*

$$\lim_{n \to \infty} w(z_{a,n}) = a.$$

Of course, the theorem of Picard also holds for (ν, μ)-solutions.

1.9 Theorem. *Let z_0 be an essential singularity of the (ν, μ)-solution $w(z)$. Then in every neighbourhood of z_0 $w(z)$ assumes every value from $\mathbb{C}$ except for, at most, one value $a \in \mathbb{C}$.*

Furthermore, also for (ν, μ)-solution we have the

1.10 Uniqueness theorem. *Let $w_1(z)$, $w_2(z)$ be (ν, μ)-solutions in the domain G (with one and the same ν, μ), and let z_j, $j = 1, 2, \ldots$, be any sequence of points from G with the set $\{z_j\}$ having at least one accumulation point in G. If $w_1(z_j) = w_2(z_j)$ $\forall j = 1, 2, \ldots$, then $w_1(z) \equiv w_2(z)$ in G.*

Proof. The function $w(z) = w_1(z) - w_2(z)$ is again a (ν, μ)-solution in G and, therefore, admits a representation $w = f \circ \chi(z)$. The zeroes of the analytic function f have at least one accumulation point in $\chi(G)$. By the uniqueness theorem for analytic functions we then have $f(\chi) \equiv 0$ in $\chi(G)$, hence $w(z) \equiv 0$ in G, which we had to prove.

III.2 ARGUMENT PRINCIPLE, INVARIANCE OF DOMAINS, MAXIMUM PRINCIPLE

2.1 Argument principle. *Let G be a domain $\subseteq \mathbb{C}$, and $\mathfrak{C}$ any positively oriented Jordan curve $\subset G$ with the interior $D \subset G$. Let $w(z)$ be a (ν, μ)-solution in G except for finitely many poles in D with total number P ($=$ sum of the orders of all poles within D), and $w(z) \neq 0$ on $\mathfrak{C}$. Then we have for the increase of the argument of $w(z)$ along $\mathfrak{C}$*

(I) $\Delta_{\mathfrak{C}} \arg w(z) = \Delta_{\mathfrak{C}} \log w(z) = N - P$

where N means the total number of the zeroes of $w(z)$ within D.

Proof. We have $\Delta_{\mathfrak{C}} \log w(z) = \Delta_{\mathfrak{C}} \log f \circ \chi(z) = \Delta_{\chi(\mathfrak{C})} \log f(\chi) = N - P$ in view of the argument principle for analytic functions and definitions 1.2, 1.5. The argument principle is proved.

As an application of the argument principle we first give a generalization of Rouché's theorem for (ν, μ)-solutions.

2.2 Theorem. *Let $w_1(z)$, $w_2(z)$ be (ν, μ)-solutions in the domain G, and let $\mathfrak{C}$ be a Jordan curve $\subset G$ with interior $D \subset G$. On $\mathfrak{C}$, let*

$$|w_1(z)| < |w_2(z)|.$$

Then $w_2(z)$ and $w_1(z) + w_2(z)$ have the same total number of zeroes within D.

Proof. We have $\Delta_{\mathfrak{C}} \log [w_1(z) + w_2(z)] = \Delta_{\mathfrak{C}} \log w_2(z) + \Delta_{\mathfrak{C}} \log \left(1 + \dfrac{w_1(z)}{w_2(z)}\right)$ $= \Delta_{\mathfrak{C}} \log w_2(z)$. By the argument principle this is our assertion.

Again by the representation theorem, and now in conjunction with the invariance of domains for analytic functions, we obtain

2.3 Theorem. (*On the invariance of domains.*) *Let $w(z)$ be a non-constant (ν, μ)-solution in the domain G. Then $w(G)$ is a domain.*

(Of course, this theorem could be extended to a stronger version.) Also an immediate consequence from the representation theorem (or now also from the theorem on the invariance of domains) is the following.

2.4 Maximum principle. *If $w(z) = u(z) + iv(z)$ is a (ν, μ)-solution in the domain $G \subseteqq \overline{\mathbb{C}}$ and if $|w|$ or u or v take a local maximum at any point $z_0 \in G$, then $w = u + iv$ is constant in G.*

Proof. Because a non-constant $w(z)$ maps a neighbourhood of $z_0 \in G$ onto a neighbourhood of the (finite) point $w(z_0)$, $|w(z_0)|$ cannot be a local maximum. In the same way, $e^{w(z)}$ and $e^{-iw(z)}$ map a neighbourhood of z_0 onto a neighbourhood of $e^{w(z_0)}$ and $e^{-iw(z_0)}$, respectively, if $w(z)$ is non-constant. Hence

$$|e^{w(z_0)}| = e^{u(z_0)} \quad \text{and} \quad |e^{-iw(z_0)}| = e^{v(z_0)}$$

cannot yield local maxima of $|e^{w(z)}| = e^{u(z)}$ and $|e^{-iw(z)}| = e^{v(z)}$, respectively. By monotonicity of the real exponential function, $u(z_0)$ and $v(z_0)$ therefore cannot be local maxima of $u(z)$ and $v(z)$, respectively, unless $w(z)$ is constant.

Together with the maximum principle there also holds the following

2.5 Minimum principle. *Let $w(z) = u(z) + iv(z)$ be a (ν, μ)-solution in the domain $G \subseteqq \overline{\mathbb{C}}$, and $w(z) \neq 0$ in G. If $|w|$ takes a local minimum at any point $z_0 \in G$, then $w(z)$ is constant in G.*

Proof. Although $h(z) := 1/w(z)$ is in general not a (ν, μ)-solution in G, it still admits a representation $h(z) = f \circ \chi(z)$, where χ is schlicht in G and f is analytic in $\chi(G)$. Hence, $|h(z)|$ cannot have a local maximum in G except when $w(z)$ is a constant. The minimum principle is proved.

Also (a generalization of) Liouville's theorem is true, namely

2.6 Theorem. (I) *Let $w(z)$ be a (ν, μ)-solution bounded in $\mathbb{C}$. Then $w(z)$ is constant in $\mathbb{C}$.*

(II) *If $w(z)$ is a (ν, μ)-solution in $\overline{\mathbb{C}}$, then $w(z)$ is constant in $\mathbb{C}$.*

(I) is just the original terminology of (II), and (II) holds on account of the maximum principle.

There is also a sharper form of Liouville's theorem 2.6, namely

2.7 Theorem. *Let $w(z)$ be a (ν, μ)-solution in $\mathbb{C}$, and for every positive γ let a constant K_γ exist such that*

(1) $|w(z)| \leqq K_\gamma |z|^\gamma$ $\forall z$ *with* $|z| > 1$.

Then $w(z)$ is constant in $\mathbb{C}$.

Proof. The function $g(t) := w(1/t)$ is a (ν^*, μ^*)-solution with certain coefficients $\nu^*(t)$, $\mu^*(t)$ in $\{0 < |t| < 1\}$. Hence $g(t)$ admits a representation $g(t) = f \circ \chi(t)$ in $\{0 < |t| < 1\}$, where $\chi(t)$ is a hydrodynamically normalized Beltrami homeomorphism of $\mathbb{C}$ onto itself, and f is analytic in $\chi(\{0 < |t| < 1\})$. Now, $\chi(t)$ is Hölder continuous at $t = 0$. By 2.7(1) (with any γ fixed) $\chi(0)$ cannot be an essential singularity, but since $\gamma > 0$ is arbitrary, it cannot be a pole, either. Hence $\chi(0)$ is a removable singularity of f, that is, $w(z)$ must be bounded in $\mathbb{C}$. By 2.6 $w(z)$ then is constant in $\mathbb{C}$, which we had to prove.

In an analogous way, by means of the representation theorem, the Hölder continuity of quasiconformal mappings and the Schwarz formula for analytic functions (cf., e.g. [B/S], p. 160, formula (8)) we obtain the following.

2.8 Theorem. *Let $w(z)$ be a (ν, μ)-solution in $\mathbb{C}$, and for every positive constant γ let a constant K_γ exist such that*

$$|\mathrm{Re}\, w(z)| \leqq K_\gamma |z|^\gamma \quad \forall z \text{ with } |z| > 1.$$

Then $w(z)$ is constant in $\mathbb{C}$.

The definitions and statements 1.2, 1.5, 1.6, 1.8, 1.9, 2.1, 2.3 to 2.6 remain valid if the condition for $w(z)$ to be a (ν, μ)-solution at a point z_0 or in a domain G is replaced by the condition for $w(z)$ to admit a representation 'analytic function $\circ$ schlicht mapping' in a neighbourhood of z_0 or in a domain G, cf. [S 7].

III.3 ASYMPTOTIC EXPANSIONS AT FINITE POINTS

According to definition 1.2, with every zero z_0 of a (ν, μ)-solution $w(z)$ we have associated a positive integer to be the order of this zero. Now there arises the question of the extent to which the order of a zero as well as the order of a pole determine the asymptotic behaviour at the zero and the pole, respectively. The following example shows that the one must have nothing to do with the other.

The function $w(z) = z \cdot |z|^{(1-K)/K}$, where K is a real constant >1, and is a schlicht solution of the Beltrami system $w_{\bar{z}} = \nu w_z$ in $\mathbb{C}$, where

3.1
$$\nu(z) = -\frac{K-1}{K+1}\frac{z}{\bar{z}}.$$

Because of $w(0) = 0$, $w(1) = 1$, this $w(z)$ coincides with $f^\nu(z)$, cf. I.2.1. Thus, $w(z)$ has the representation $f \circ \chi(z)$, where $\chi(z) \equiv f^\nu(z)$ and $f(\chi) \equiv \chi$. Hence, z_0 is a zero of order one for every $K > 1$. On the other hand, we have

$|w(z)| = |z|^{1/K}$. That means that, if we do not put some additional conditions on ν, μ, there need not exist any asymptotic expansions of the form

$$w(z) = az + b\bar{z} + o(z)$$

at zeroes of order one, as were desirable.

We have the following

3.2 Theorem. *Let G be a domain $\subseteqq \mathbb{C}$, $z_0 \in G$, $\nu, \mu \in HL_p(\{z_0\}, G)$, $2 < p$ and $|\nu(z)| + |\mu(z)| \leqq$ constant $=: k < 1$ $\forall z \in G$.*

(I) *Every (ν, μ)-solution $w(z)$ in G with a zero of order n at z_0 admits an asymptotic expansion*

$$w(z) = c_1[(z - z_0) + \ell(\overline{z - z_0})]^n - \overline{c_1} b[\overline{z - z_0} + \bar{\ell}(z - z_0)]^n + O(|z - z_0|^{n+\alpha})$$

with certain constants c_1, b, ℓ and $\alpha > 0$.

(II) *Every (ν, μ)-solution $w(z)$ in $G \setminus \{z_0\}$ with a pole of order m at z_0 admits an asymptotic expansion*

$$w(z) = \left[\frac{c_2}{(z - z_0 + \ell(\overline{z - z_0}))^m} - \frac{b\overline{c_2}}{(\overline{z - z_0} + \bar{\ell}(z - z_0))^m}\right](1 + O(|z - z_0|^{\alpha}))$$

with certain constants c_2, b, ℓ and $\alpha > 0$.

(III) *For the constants mentioned under* (I) *and* (II) *it holds*

$$c_1 \neq 0, \qquad c_2 \neq 0,$$

$$b = 2\varkappa/(1 + \sqrt{1 - 4|\varkappa|^2}) =: h(\varkappa), \qquad \varkappa = -\mu(z_0)/(1 + |\mu(z_0)|^2 - |\nu(z_0)|^2),$$

$$\ell = h(\sigma), \qquad \sigma = \nu(z_0)/(1 + |\nu(z_0)|^2 - |\mu(z_0)|^2);$$

α can be chosen from (0, 1) *in such a way that it depends only on k.*

(IV) *If, in addition, ν, μ satisfy*

(1) $\lim_{z \to z_0} \nu(z) = \nu(z_0), \quad \lim_{z \to z_0} \mu(z) = \mu(z_0),$

then we may put $\alpha = 1 - (2/p)$ with $p > 2$ mentioned above.

Proof. By means of the affine transformation

3.3 $$h = w + b\bar{w}$$

with b from 3.2(III), $w(z)$ changes into a (ν_0, μ_0)-solution $h(z)$ in G (in $G \setminus \{z_0\}$) where ν_0, μ_0 are those coefficients mentioned in II.3.34, and $\mu_0(z_0) = 0$. By the argument principle $h(z)$ has a zero of order n (a pole of order m) at z_0 if and only if $w(z)$ has a zero of order n (a pole of order m) at z_0. Together with ν, μ, also $\nu_0, \mu_0 \in HL_p(\{z_0\}, G)$. The function $h(z)$ admits a representation $h(z) = f \circ \chi(z)$ where $\chi(z) = f^{\nu^*}(z)$ (see I.2.1) with

3.4 $$\nu^* = \nu_0(z) + \mu_0(z)\frac{\overline{h_z}}{h_z} \text{ as } z \in G, \qquad \nu^*(z) = 0 \text{ otherwise,}$$

and f is an analytic function possessing either a zero of order n or a pole of order m at $\chi(z_0)$.

Now, because of $\mu_0(z_0) = 0$, together with ν_0, μ_0, then $\nu^* \in HL_p(\{z_0\}, G)$. Hence, by theorem II.5.2 (cf. also II.5.22, where $w(z)$ is to be replaced by $\chi(z)$), we have

3.5 $\chi(z) = \chi(z_0) + \chi_z(z_0)(z - z_0) + \chi_{\bar{z}}(z_0)(\overline{z - z_0}) + O(|z - z_0|^{1+\alpha})$,

with $\alpha \in (0, 1)$ depending only on k (cf. the remarks following II.5.11) and $\chi_z(z_0) \neq 0$ (cf. theorem II.5.47). Furthermore,

3.6 $$h(z) = (1/n!) f^{(n)}(\chi_0)(\chi - \chi_0)^n + O(|\chi - \chi_0|^{n+1}),$$

where $\chi = \chi(z)$, $\chi_0 = \chi(z_0)$, $f^{(n)}(\chi_0) \neq 0$, or

3.7 $$h(z) = \frac{a_{-m}}{(\chi - \chi_0)^m}(1 + O(\chi - \chi_0)), \quad a_{-m} \neq 0,$$

if z_0 is a zero of order n or a pole of order m, respectively, of h. Inserting 3.5 into 3.6, 3.7, we obtain

3.8 $$h(z) = A_1(z - z_0 + b(\overline{z - z_0}))^n + O(|z - z_0|^{n+\alpha})$$

$$\text{with } b = \frac{\chi_{\bar{z}}(z_0)}{\chi_z(z_0)} = \nu_0(z_0)$$

(cf. II.3.31, II.3.38), respectively

3.9 $$h(z) = A_2(z - z_0 + b(\overline{z - z_0}))^{-m}(1 + O(|z - z_0|^\alpha)),$$

if z_0 is a zero of order n, respectively a pole of order m. Here A_1 and A_2 are (finite) constants different from zero. From 3.3, 3.8, 3.9 we obtain the expansions for $w(z)$ as asserted in (I), (II), (III).

In addition, let ν and μ now satisfy the conditions of (IV). Let us assume first that $\nu(z_0) = 0$; otherwise we only have to put the z-plane through a suitable affine mapping, cf. II.3.28, II.3.31. Further we may restrict G in such a way that, by the continuity of ν, μ at z_0 and $\mu_0(z_0) = 0$, $\nu(z_0) = \nu_0(z_0) = 0$, we have $k'C(p) < 1$ with $k' := \|\nu^*\|_{L_\infty} = \|\nu^*\|_{L_\infty(G)}$. This means that we may replace p' by p in II.5.11, II.5.22 (with $w(z)$ replaced by $\chi(z)$) and, consequently, we may put $\alpha = 1 - (2/p)$ in 3.5 and in all the other relations above resting upon 3.5. Reversing the initially mentioned affine mapping of the z-plane (to ensure $\nu(z_0) = 0$), we obtain asymptotic expansions as in (I) and (II) with $\alpha = 1 - (2/p)$ and the same constants as in (I) to (III) (cf. section II.3 after II.3.36). The latter also follows in a direct way, namely it holds:

3.10 *The coefficients* c_1, c_2, b, b *of the asymptotic expansions* 3.2(I), (II) *are uniquely determined.*

Since, let $z_0 = 0$, e.g., and

3.11 $$w(z) = \left[\frac{c}{(z + b\bar{z})^m} - \frac{b\bar{c}}{(\bar{z} + \bar{b}z)^m}\right](1 + O(|z|^\alpha))$$

$$= \left[\frac{c'}{(z + b'\bar{z})^m} - \frac{b'\overline{c'}}{(\bar{z} + \overline{b'}z)^m}\right](1 + O|z|^{\alpha'})).$$

Putting $z = t\,e^{is}$, multiplying both sides by t^m and, if this is done, letting t tend to zero along the positively real axis, we obtain, on putting again $e^{is} = z$,

3.12 $$\frac{c}{(z^2+\ell)^m} - \frac{b\bar{c}}{(\bar{\ell}z^2+1)^m} = \frac{c'}{(z^2+\ell')^m} - \frac{\overline{b'c'}}{(\overline{\ell'}z^2+1)^m} \quad \text{for } |z| = 1.$$

By the uniqueness theorem for analytic functions, 3.12 must then hold for all z different from $\pm i\sqrt{\ell}$, $\pm i\sqrt{\ell'}$. In any case, both the absolute values of ℓ and ℓ' are less than 1. Hence $\ell = \ell'$ and, consequently, $c = c'$, $b = b'$. In a corresponding way we conclude that the coefficients in an asymptotic expansion (I) are also uniquely determined. This proves both 3.10 and 3.2(IV). Theorem 3.2 is proved.

If z_0 is both a zero of order n of the (ν, μ)-solution $w_1(z)$ and a zero of order m of the (ν, μ)-solution $w_2(z)$, then z_0 is a zero of a certain order l of $w(z) = w_1(z) + w_2(z)$. If the conditions of theorem 3.2 are satisfied, then by 3.2 we have

3.13 $$l \geqq \min\{n, m\}$$

(a corresponding relation holds for poles). It is an open and, possibly, important question of whether 3.13 (and the corresponding relation for poles) also holds without any additional smoothness condition on ν, μ (as is $\nu, \mu \in HL_p(\{z_0\}, G)$).

III.4 ASYMPTOTIC EXPANSIONS AT THE POINT ∞, COMPACTNESS

In dealing with asymptotic expansions at ∞ it is useful to also consider other types of smoothness conditions on ν, μ.

4.1 *We will say that ν, μ, defined and measurable in $\bar{\mathbb{C}}$, satisfy the condition (L^q) if*

(1) $|\nu(z) - \nu(\infty)| + |\mu(z) - \mu(\infty)| \in L_q$ *with* $1 \leqq q < 2$

and

(2) $|\nu(z)| + |\mu(z)| \leqq$ *constant* $=: k < 1$ $\forall z \in \mathbb{C}$.

Of course, then we also have $|\nu(\infty)| + |\mu(\infty)| \leqq k$. Condition (L^q) is the stronger the smaller q $(\geqq 1)$ is. Indeed, we have:

4.2 *If $\nu(z) - \nu(\infty) \in L_q$ with $q \geqq 1$ and $|\nu(z)| \leqq k$ in $\bar{\mathbb{C}}$, then $\nu(z) - \nu(\infty) \in L_s$ and $\|\nu(z) - \nu(\infty)\|_{L_q} \geqq \|\nu(z) - \nu(\infty)\|_{L_s}$ $\forall s \geqq q$.*

Since
$$|\nu(z) - \nu(\infty)|^q/(2k)^q \geqq |\nu(z) - \nu(\infty)|^s/(2k)^s \quad \text{if } s \geqq q \geqq 1.$$

For instance, condition (L^q) is satisfied if 4.1(2) and

4.3 $$|\nu(z) - \nu(\infty)| + |\mu(z) - \mu(\infty)| \leqq \text{constant} \cdot |z|^{-\gamma} \quad \text{for } |z| > 1$$

hold with γ to be any constant > 1.

4.4 Theorem. *Let $G \subseteq \overline{\mathbb{C}}$ be a domain, $\{|z| > R\} \subset G$, $\infty \in G$. Let Q be a set of quasiconformal mappings g of G satisfying $g(\infty) = \infty$, $|g_{\bar z}/g_z| \leqq$ constant $= k < 1$ a.e. in G, $g_{\bar z}/g_z \in L_q$, $\|g_{\bar z}/g_z\|_{L_q} \leqq$ constant $=: K$, $1 \leqq q < 2$, where k, K, q are fixed constants independent of $g \in Q$. Furthermore, let Z_1 be a bounded set $\subset G$.*

Then there exists a constant $c_1 > 0$ depending only on k, K, q, R and $\sup\{|z| : z \in Z_1\}$ and, for each $g \in Q$, a further constant $c_g \neq 0$ such that

$$\left|\frac{g(z) - g(z_1)}{z} - c_g\right| \leqq |c_g| \cdot \frac{c_1}{|z|} \quad \forall g \in Q,\ \forall z_1 \in Z_1,\ \forall z \in G \setminus \{0\}.$$

Proof. Let us fix a $\delta \in (0, 1)$ so that

4.5 $$kC(s) \leqq \frac{1+k}{2} =: k_0 < 1 \quad \forall s \in [2 - \delta, 2 + \delta]$$

(cf. 0.4.17, 0.4.18). Further let us fix $p \in (0, 2 + \delta]$ and $q \in [2 - \delta, 2)$, and, without loss in generality, let this q coincide with that mentioned in 4.4. For $g \in Q$ we define

4.6 $\nu_g(z) = g_{\bar z}(z)/g_z(z)$ for almost every $z \in G$, $\nu_g(z) = 0$ elsewhere in $\mathbb{C}$.

By 4.2 this $\nu_g(z)$ belongs to $L^{p,q}$, and we have $\|\tilde\nu_g\|_{L^{p,q}} \leqq K$. The equation

4.7 $$h = \nu_g + \nu_g Th, \quad \text{where} \quad \|\nu_g Th\|_{L^{p,q}} \leqq k_0 \|h\|_{L^{p,q}},$$

has just one solution $h \in L^{p,q}$, by the Banach fixed point theorem, and

4.8 $$\|h\|_{L^{p,q}} \leqq (1 - k_0)^{-1} \|\nu_g\|_{L^{p,q}} \leqq (1 - k_0)^{-1} K =: K^*.$$

Then

4.9 $$H_g(z) := z + Ph(z)$$

is a solution of the $(\nu_g, 0)$-system in $\mathbb{C}$, by 0.4.19 and 4.7. Hence we have

4.10 $$H_g(z) = F \circ f^{\nu_g}(z)$$

(cf. I.2.1, I.1.7(II)), where F is analytic in $\mathbb{C}$. By 0.4.3(IV) $H_g(z) \to \infty$ as $z \to \infty$ and $\Delta_{\mathfrak{C}} \log \dfrac{H_g(z)}{z} = 0$ if $\mathfrak{C}$ is a sufficiently large circle with centre at the origin, $\mathfrak{C}$ positively oriented. Thus, $\Delta_{\mathfrak{C}} \log H_g(z) = 1$ and, therefore, $\Delta_{f^{\nu_g}(\mathfrak{C})} \log F(\chi) = 1$. The latter means that F has a pole of order 1 at ∞ and, consequently, is an entire linear transformation. In particular, H_g is schlicht in $\mathbb{C}$. Furthermore, by 0.4.3(I), 4.8, 4.9, we have

4.11 $\{H : H = H_g(z), |z| > R\} > \{|H| > R^*\}$ where $R^* = R + K_{p,q}K^*$.

Since both g and H_g are solutions of the same Beltrami system in $G \setminus \{\infty\}$, there holds a relation

4.12 $$g(z) = \zeta_g \circ H_g(z)$$

where ζ_g is analytic and schlicht in $H_g(G) \setminus \{\infty\}$ and possesses, consequently, a Laurent expansion

4.13 $$\zeta_g(H) = c_g H + a_0 + \frac{a_1}{H} + \ldots \quad \text{as } |H| > R^*, \text{ where } c_g \neq 0$$

(of course, the a_j, $j = 1, 2, \ldots$, also depend on the g at the time, from Q). The coefficients a_j/c_g, $j = 1, 2, \ldots$, of the hydrodynamically normalized conformal mappings

4.14 $$(\zeta_g(H) - a_0)/c_g = H + \frac{a_1/c_g}{H} + \frac{a_2/c_g}{H^2} + \ldots$$

of the domain $\{|H| > R^*\}$ satisfy

4.15 $$|a_j/c_g| \leqq j^{-1/2}(R^*)^{j+1}, \quad j = 1, 2, \ldots,$$

by the Bieberbach area theorem (cf., e.g. [B3], p. 99).

Now let $R_0 = \max\{3R^*, \sup\{|z| : z \in Z_1\}\}$. Because $|H_g(z)| \geqq 2R^*$ if $|z| \geqq R_0$ and by 4.15 we have

4.16

$$\left|\frac{g(z) - a_0}{z} - c_g\right| \leqq \left|\frac{c_g}{z}\right| \cdot \left|Ph(z) + \frac{a_1/c_g}{H_g(z)} + \ldots\right| \leqq \left|\frac{c_g}{z}\right| R_0 \quad \text{for } |z| \geqq R_0.$$

Hence

4.17 $$|g(z) - a_0| \leqq |c_g| \cdot 2R_0 \quad \text{if } |z| = R_0.$$

Since $g(z) - a_0$ is a schlicht mapping of G with $\infty \to \infty$, 4.17 also holds for all $z \in G$ with $|z| \leqq R_0$, and, in particular, also for all $z \in Z_1$. From 4.16 and 4.17 we obtain

4.18 $$\left|\frac{g(z) - g(z_1)}{z} - c_g\right| \leqq |c_g| \frac{5R_0}{|z|},$$

which proves 4.4.

4.19 *From now on whenever (in connection with classes of (ν, μ)-solutions or quasiconformal mappings) we speak about convergence in a domain $G \subseteq \overline{\mathbb{C}}$ and compactness in G, then this always means locally uniform convergence in $G \setminus \{\infty\}$ and compactness with respect to locally uniform convergence in $G \setminus \{\infty\}$, respectively.*

4.20 Theorem. *Let $\mathfrak{P}$ be a set of pairs (ν, μ) with the following properties:*

(a) *ν, μ are functions defined and measurable in $\overline{\mathbb{C}}$ and $|\nu(z)| + |\mu(z)| \leqq k < 1$ $\forall z \in \mathbb{C}$,*

(b) *$\nu(\infty) = A$, $\mu(\infty) = B$,*

(c) *$\|\nu - A\|_{L_q} + \|\mu - B\|_{L_q} \leqq K$, $1 \leqq q < 2$,*

where k, A, B, K, q are fixed constants independent of $(\nu, \mu) \in \mathfrak{P}$.

Let G be a domain as in 4.4, P be a finite point from G. Let $\mathfrak{Q}(P)$ be a non-empty set of schlicht mappings f of G with the following properties:

(α) *For every $f \in \mathfrak{Q}(P)$ there exists (at least) one $(\nu, \mu) \in \mathfrak{P}$ so that f is a (ν, μ)-solution in $G \setminus \{\infty\}$,*

(β) *$f(\infty) = \infty$, $f(P) = P$,*

(γ) *$f(z) = z + (A + B)\bar{z} + \varepsilon_f(z)$ where $\lim_{z \to \infty} \frac{\varepsilon_f(z)}{z} = 0$.*

Then it holds:

(I) *There exists a constant c_2 depending only on k, K, G, $|A|$, $|B|$, q and P so that $|\varepsilon_f(z)| \leqq c_2$ for all $f \in \mathfrak{Q}(P)$ and all $z \in G \setminus \{\infty\}$.*

(II) *For every $f \in \mathfrak{Q}(P)$ there exists the limit*

$$\lim_{z \to \infty} \varepsilon_f(z) =: a(f).$$

(III) *The sets $\mathfrak{Q}(P)$ and $\mathfrak{Q} = \{f(z) - a(f) : f \in \mathfrak{Q}(P)\}$ are relatively compact in the set of the $(1 + k)/(1 - k)$-quasiconformal mappings of G possessing an expansion (γ) satisfying the estimate mentioned under* (I).

Proof. The affine transformation

4.21 $$t = z + \ell(\overline{z - P}) \quad \text{with } \ell = h(\sigma), \quad \sigma = \frac{A}{1 + |A|^2 - |B|^2}$$

(note that $|A| + |B| < 1$, by 4.20(a), (b), (c)) carries $f \in \mathfrak{Q}(P)$ into $f_1 \in \mathfrak{Q}_1(P) := \{f_1(t) = f(z(t)) : f \in \mathfrak{Q}(P)\}$, where $z(t)$ means the inverse mapping of 4.21. This $f_1(t)$ is a solution of

$$f_{1\bar{t}} = \nu_1(t) f_{1t} + \mu_1(t) \overline{f_{1t}}$$

in the image domain G_t of G under the mapping 4.21, if f is a (ν, μ)-solution in G and ν_1, μ_1 correspond to ν, μ according to II.3.30. Instead of the pairs $(\nu, \mu) \in \mathfrak{P}$ we now consider the set of pairs $\mathfrak{P}_1 \ni (\nu_1, \mu_1)$ where the pairs (ν, μ) and (ν_1, μ_1) are to correspond according to 4.21, II.3.30. For $(\nu_1, \mu_1) \in \mathfrak{P}_1$ we then have

4.22

$$|\nu_1(t)| + |\mu_1(t)| \leqq k_1 < 1 \quad \forall z \in \overline{\mathbb{C}} \left(\frac{1 + k_1}{1 - k_1} \leqq \frac{1 + k}{1 - k} \frac{1 + |\ell|}{1 - |\ell|}, \quad \text{cf. II.3} \right),$$

$$\nu_1(\infty) = 0, \qquad \mu_1(\infty) = B_1, \qquad \|\nu_1\|_{L_q} + \|\mu_1 - B_1\|_{L_q} \leqq K_1,$$

where k_1 depends only on $|A|$, $|B|$ and k, B_1 depends only on A and B and K_1 depends only on k, $|A|$, $|B|$, K and q.

Furthermore, by means of the affinity

4.23 $$g = f_1 + b(\overline{f_1 - P}) \quad \text{with } b = h(\varkappa), \quad \varkappa = \frac{-B}{1 + |B|^2 - |A|^2},$$

of the f_1-plane we obtain a set $\mathfrak{P}_2$ of pairs (ν_2, μ_2) and a set of mappings $\mathfrak{Q}_2(P) = \{g(t) = f_1(t) + b(\overline{f_1(t) - P}) : f_1 \in \mathfrak{Q}_1(P)\}$ from $\mathfrak{P}_1$ and $\mathfrak{Q}_1(P)$, respectively, according to II.3.33 to 36, where

4.24

$$|\nu_2(t)| + |\mu_2(t)| \leqq k_2 < 1 \quad \forall z \in \overline{\mathbb{C}} \left(\frac{1 + k_2}{1 - k_2} \leqq \frac{1 + k_1}{1 - k_1} \frac{1 + |b|}{1 - |b|} \right),$$

$$\nu_2(\infty) = \mu_2(\infty) = 0, \qquad \|\nu_2\|_{L_q} + \|\mu_2\|_{L_q} \leqq K_2.$$

Here k_2 depends only on $|A|$, $|B|$ and k, and K_2 depends only on k, $|A|$, $|B|$, K and q.

$\mathfrak{Q}_2(P)$ satisfies the conditions of theorem 4.4 with $Z_1 = \{P\}$. Hence for all $g \in \mathfrak{Q}_2(P)$ we have

$$\left|\frac{g(t) - P}{t} - c_g\right| \leqq |c_g| \cdot c_1 \cdot |t|^{-1} \quad \forall t \in G_t \setminus \{0\}. \tag{4.25}$$

Now, because the transformations 4.21 and 4.23 do not affect the property to admit an expansion of the form 4.20(γ), we have

$$g(t) = A^*t + B^*\bar{t} + \varepsilon_g^*(t), \qquad \lim_{t\to\infty} \frac{\varepsilon_g^*(t)}{t} = 0, \tag{4.26}$$

where A^*, B^* are constants depending only on A and B. On comparing 4.25 and 4.26 we obtain $c_g = A^*$, $B^* = 0$ $\forall g \in \mathfrak{Q}_2(P)$ as well as $|\varepsilon_g^*(t)| \leqq |A^*|\, c_1 + |P|$. This gives (I) after reversing the affine mappings 4.21 and 4.23.

Every $g(t) \in \mathfrak{Q}_2(P)$ admits a representation

$$g(t) = A^* H_g(t) + a_0 + \frac{a_1}{H_g(t)} + \frac{a_2}{(H_g(t))^2} + \ldots \tag{4.27}$$

where $H_g(t) = t + Ph(t)$, cf. 4.6, 4.7, 4.12, 4.13. By 0.4.3(IV) $\lim_{t\to\infty} \varepsilon_g^*(t) = a_0$. Since $\varepsilon_f(z(t))$ arises from $\varepsilon_g^*(t)$ by means of an affine transformation, this proves (II).

By (II) the mappings from $\mathfrak{Q}(P)$ are uniformly bounded in bounded subdomains of G. By theorem II.4.1 every sequence $w_n \in \mathfrak{Q}(P)$ contains a convergent subsequence such that

$$F_n \circ \chi_n(z) = w_n(z) \to F \circ \chi(z) = w(z),$$

where χ is a $(1 + k)/(1 - k)$-quasiconformal mapping of $\mathbb{C}$ onto itself and F is analytic in $\chi(G \setminus \{\infty\})$. Here χ is the locally uniform limit of the χ_n, and F is the limit of the conformal mappings F_n, where the convergence is locally uniform in $\chi(G \setminus \{\infty\})$. By 4.20(γ) F cannot be constant, hence it is schlicht (cf., e.g. [B3], p. 65). Consequently, w is a quasiconformal mapping of $G \setminus \{\infty\}$ and, therefore, also of G. Because $w_n(z) = z + (A + B)\,\bar{z} + \varepsilon_n(z)$ converges locally uniformly in $G \setminus \{\infty\}$, then $\varepsilon_n(z)$ also converges locally uniformly in $G \setminus \{\infty\}$ to a bounded function $\varepsilon(z)$. In an analogous way we deal with $\mathfrak{Q}$. Assertion (III) is proved.

Of course, theorem 4.20 requires a completion in view of existence of such non-empty classes $\mathfrak{Q}(P)$. This is a consequence of the following.

4.28 Theorem. *Let G, $\mathfrak{P}$ be as in* 4.20, *let $\mathfrak{P}$ be non-empty, and let (ν, μ) be an arbitrary but fixed pair from $\mathfrak{P}$. There exist schlicht (ν, μ)-solutions $f(z)$ in $G \setminus \{\infty\}$ with an asymptotic expansion*

(I) $f(z) = cz + d\bar{z} + \varepsilon_f(z)$ *where* $c \neq 0$ *and* $\lim_{z\to\infty} \dfrac{\varepsilon_f(z)}{z} = 0$, *if and only if* $d = cA + \bar{c}B$.

In particular, for every $c \neq 0$ there exists a schlicht (ν, μ)-solution $f_c(z)$ in $\mathbb{C}$ such that

(II) $f_c(z) = cz + (cA + \bar{c}B)\,\bar{z} + \varepsilon(z)$ *where* $\varepsilon(z) \to 0$ *as* $z \to \infty$.

Proof. To start with let $d = cA + \bar{c}B$. We ask for a schlicht (ν, μ)-solution $f_c(z)$ in $\mathbb{C}$ having an expansion 4.28 (II). To this end, we put

4.29 $$f_c(z) = cz + (cA + \bar{c}B)\,\bar{z} + Ph(z),$$

while hoping that this device leads to an $h \in L^{p,q}$ with $1 < q < 2 < p < \infty$. According to 0.4.19 for h we then obtain the equation

4.30 $$h = (\nu - A)\,c + (\mu - B)\,\bar{c} + \nu Th + \mu\overline{Th}.$$

Because

$$\|\nu Th + \mu\overline{Th}\|_{L^{p,q}} \leqq k \cdot \max\{C(p), C(q)\} \cdot \|h\|_{L^{p,q}} \leqq k_0\|h\|_{L^{p,q}}$$

for some p, q satisfying $1 < q < 2 < p < \infty$ (cf. 4.5), equation 4.30 has a unique solution by the Banach fixed point theorem (while $\nu Th + \mu\overline{Th}$ is not complex-linear, it is additive). Then the f_c constructed with $h \in L^{p,q}$ satisfying 4.30 will become a (ν, μ)-solution in $\mathbb{C}$ having the desired asymptotic behaviour as $z \to \infty$ (cf. 0.4.3 (IV)). From the representation theorem and the argument principle then the schlichtness of f_c also follows; cf., for example, the proof of the schlichtness of H_g in 4.9, 4.10 above, note $|A| + |B| < 1$.

Now let $d \neq cA + \bar{c}B$ and $f(z)$ be any (ν, μ)-solution with expansion 4.28 (I). Then, also, $g(z) := f(z) - f_c(z)$ is a (ν, μ)-solution in $G \setminus \{\infty\}$ and

4.31

$$g(z) = (d - cA - \bar{c}B)\,\bar{z} + \varepsilon_g(z), \qquad \lim_{z \to \infty} \frac{\varepsilon_g(z)}{z} = 0, \qquad g(z) = F \circ \chi(z),$$

where χ is a quasiconformal mapping of $\mathbb{C}$ and F is analytic in $\chi(G \setminus \{\infty\})$. The argument principle then gives a contradiction with $d - cA - \bar{c}B \neq 0$. Theorem 4.28 is proved.

4.32 ***Remark.*** The mapping f_c from 4.28 (II) also exists under weaker assumptions on ν, μ, for example,

$$\nu - \nu(\infty), \quad \mu - \mu(\infty) \in L_p, \qquad 2 < p < \infty,$$

(cf. [R4]), and, of course, theorem 4.28 could also be formulated with respect to a finite point, in place of the point of infinity.

Now let $f(z)$ be a (ν, μ)-solution in $G \setminus \{\infty\}$, let G be a domain containing ∞, and let ∞ be a pole of order m of $f(z)$. We again assume

4.33 $$\nu - \nu(\infty), \quad \mu - \mu(\infty) \in L_q, \qquad 1 \leqq q < 2.$$

By means of suitable affine transformations of the z- and f-planes (cf. 4.21 to 4.24), from ν, μ we obtain $\nu_2, \mu_2 \in L_q$, and the corresponding $g(t)$ (cf. the proof of 4.20) again has a pole of order m at ∞, by the argument principle. By the representation theorem, 4.4 and 4.28, we have

4.34 $$g(t) = a_m(t + \varepsilon(t))^m + \ldots + a_1(t + \varepsilon(t)) + a_0 + b_1(t + \varepsilon(t))^{-1} + \ldots$$

where $\varepsilon(t) \to 0$ as $t \to \infty$, thus

$$g(t) = a_m \cdot t^m + t^{m-1}O(1).$$

Having reversed the affine mappings, for $f(z)$ we obtain an asymptotic expansion

4.35 $$f(z) = c(z + \ell\bar{z})^m - \bar{c}b(\bar{z} + \bar{\ell}z)^m + z^{m-1}O(1)$$

with ℓ, b as in 3.2(III), where we only have to replace $\nu(z_0)$, $\mu(z_0)$ by $\nu(\infty)$, $\mu(\infty)$, respectively.

In a corresponding way we obtain an asymptotic expansion

4.36 $$f(z) = \left[\frac{c}{(z + \ell\bar{z})^n} - \frac{b\bar{c}}{(\bar{z} + \bar{\ell}z)^n}\right](1 + |z|^{-1}O(1)),$$

with the same ℓ, b as in 4.35, if ∞ is a zero of order n of the (ν, μ)-solution f with ν, μ satisfying 4.33.

By means of affine transformations and a following inversion (of the t-plane), from 4.35, 4.36 we obtain

4.37 Theorem. *If* $(\nu(z) - \nu(z_0))/|z - z_0|^{4/q}$ *and* $(\mu(z) - \mu(z_0))/|z - z_0|^{4/q}$ *belong to* L_q, $1 \leq q < 2$, *then every* (ν, μ)*-solution having a zero of order* n *or a pole of order* m *at* z_0 *admits an expansion* 3.2(I) *or* 3.2(II), *respectively, with* $\alpha = 1$.

Proof. Let $z_0 = 0$, $\nu(0) = \mu(0) = 0$ (which can be attained by means of affine transformations). Then $\tilde{w}(z) := w(1/z)$ becomes a $(\tilde{\nu}, \tilde{\mu})$-solution in a neighbourhood of ∞, where

4.38 $$\tilde{\nu}(z) = \nu\left(\frac{1}{z}\right) \cdot \frac{z^2}{\bar{z}^2}, \qquad \tilde{\mu}(z) = \mu\left(\frac{1}{z}\right), \quad \tilde{\nu}, \tilde{\mu} \in L_q.$$

From 4.35, 4.36 (with $\ell = b = 0$), applied to $\tilde{w}(z)$, our assertion follows (note also 3.10).

The conditions of theorem 4.37 are satisfied if, for instance, at the point z_0 there holds

4.39 $$|\nu(z) - \nu(z_0)| + |\mu(z) - \mu(z_0)| \leq \text{constant} \cdot |z - z_0|^\gamma \quad \text{with } \gamma > 1.$$

Of course, theorem 3.2 also has an immediate counterpart in case $z_0 = \infty$, that is, there also exist, under weaker assumptions on ν, μ at ∞ (as is 4.33), asymptotic expansions as in 3.2(I), (II). For this we first have to extend the definition of $HL_p(D, G)$ to the case $\infty \in D$. Here we restrict ourselves to $G = \mathbb{C}$.

4.40 ***Definition.*** (I) We will say that $f \in HL_p(\{\infty\})$ if and only if f is defined everywhere in $\overline{\mathbb{C}}$ and $f(1/z) - f(\infty) \in HL_p(\{0\})$. (II) Let D be any subset of $\overline{\mathbb{C}}$, $\infty \in D$, $D \neq \{\infty\}$. We will say that $f \in HL_p(D)$ if and only if $f \in HL_p(D \setminus \{\infty\}) \cap HL_p(\{\infty\})$.

Then we have the following completion of theorem 3.2.

4.41 Theorem. *Let* ∞ *be a zero of order* n *or a pole of order* $m = -n$ *of the* (ν, μ)*-solution* $w(z)$, $\nu, \mu \in HL_p(\{\infty\})$, $p > 2$. *Then* $w(z)$ *admits the asymptotic expansion*

$$w(z) = c(z + \ell\bar{z})^{-n} - b\bar{c}(\bar{z} + \bar{\ell}z)^{-n} + O(|z|^{-n-\alpha})$$

at the point of infinity, where c, α, ℓ, b *are constants as in* 3.2(III) *with* z_0 *replaced by* ∞.

Proof. Having reduced ν, μ, by means of affine transformations, to such ones with $\nu(\infty) = \mu(\infty) = 0$, we only have to carry out an inversion of the (transformed) z-plane and to apply theorem 3.2, which gives theorem 4.41.

Chapter IV

INTEGRAL TRANSFORMATIONS AND FUNDAMENTAL SOLUTIONS

In dealing with extremal problems for schlicht solutions of elliptic systems of partial differential equations, in [R2] and [R3] (cf. also chapter VI below) there arose, by necessity, transformations carrying analytic functions into (ν, μ)-solutions with $\nu = -\bar{\mu}$, where these transformations are represented by integral operators, under certain conditions on μ. Of course, this gave rise to the question of whether such integral operators also exist for arbitrary coefficients ν, μ. Indeed this is true for ν, μ having bounded support, as we will show in the first section of this chapter (cf. [R3], chap. 11, [R5]). These integral operators, and their kernels, derive much of their importance from their connection with the fundamental solutions of the adjoint $(\nu, \bar{\mu})$-system. This fact will lead us to both a very general and simple definition and a 'simple' construction of the fundamental solutions of (ν, μ)-systems ('simple' on the basis of the Calderón-Zygmund results on singular integral operators). This will be carried through in sections IV.3 and IV.4.

The integral transformations of analytic functions into (ν, μ)-solutions, derived in IV.1, can be seen as further representation formulae for (ν, μ)-solutions. On account of their 'naturality', indicated by their connections with extremal problems, one may expect that they will prove useful in other connections, too.

IV.1 INTEGRAL OPERATORS FOR (ν, μ)-SYSTEMS

Let D be a domain $\subseteq \mathbb{C}$. We put

1.1 $B_p(D) = W^1_{p,\mathrm{loc}}(D) \cap \mathscr{C}(D)$, $A(D) = \{f: f$ is analytic in $D\}$, $\mathfrak{A}(D, \nu, \mu) = \{f: f$ is (ν, μ)-solution in $D\}$.

Let the functions ν, μ be defined and measurable all over $\mathbb{C}$ (without loss of generality), and, until further notice, (namely until IV.3, inclusively) we assume

1.2 $\operatorname{supp}(|\nu| + |\mu|) =: \mathfrak{R} \Subset \mathbb{C}, \quad |\nu(z)| + |\mu(z)| \leqq \text{constant} =: k < 1 \quad \forall z \in \mathbb{C}.$

1.3 Lemma. *We choose a constant* $\delta \in (0, 1)$ *so that*

(1) $kC(s) < 1 \quad \forall s \in [2 - \delta, 2 + \delta]$.

Then the operator $V = V_{\nu,\mu}$ *defined by*

(I) $Vh(z) := h(z) - P(\nu h_z + \mu \overline{h_z})(z)$

maps each $B_p(D)$, *with* $p \in (2, 2 + \delta]$ *and* $D \Supset \mathfrak{R}$, *in a one-to-one manner onto itself (where we in* (I) *put* $\nu(z) f(z) + \mu(z) f(z) = 0$ *for any* $z \notin \mathfrak{R}$ *and any* f).

Proof. Let $h \in B_p(D)$. Then $Vh \in B_p(D)$ by 0.4.3, 0.4.19 (because of the boundedness of $\mathfrak{R}$ $\nu f + \mu \bar{f} \in L^{p,q}$ $\forall q \in [1, p]$ if $f \in L_p(\mathfrak{R})$). Conversely, for each $h \in B_p(D)$ there is just one $g \in B_p(D)$ such that $Vg = h$. Since, by 0.4.19, from $Vg = h$ we obtain the equation

1.4 $$h_z = g_z - T_{\nu,\mu}(g_z), \quad \text{where } T_{\nu,\mu} f := T(\nu f + \mu \bar{f}).$$

Now we have

1.5 $$\|T_{\nu,\mu} f\|_{L_p(\mathfrak{R})} \leqq kC(p)\, \|f\|_{L_p(\mathfrak{R})},$$

and $T_{\nu,\mu}$ is additive, Since $h_z \in L_p(D')$ for every $D' \Subset D$, the equation

1.6 $$f = h_z + T_{\nu,\mu} f$$

has exactly one solution f in $L_p(\mathfrak{R})$, $p \in (2, 2 + \delta]$, by the Banach fixed point theorem. With this f we put

1.7 $$g := h + P(\nu f + \mu \bar{f}).$$

Of course, $g \in B_p(D)$. Further, by 0.4.19 and 1.6 $\nu g_z + \mu \overline{g_z} = \nu f + \mu \bar{f}$, thus

1.8 $$Vg = g - P(\nu f + \mu \bar{f}) = h.$$

There is no other $g \in B_p(D)$ satisfying $Vg = h$ than that defined in 1.7. For let $g_1, g_2 \in B_p(D)$ with $Vg_1 = Vg_2 = h$, that is, $V(g_1 - g_2) = 0$ in D. By 1.5 and 1.6 we first obtain $g_{1z} = g_{2z}$ a.e. in $\mathfrak{R}$. From the definition 1.3(I) of V it then finally follows that $g_1 = g_2$ in D, which completes the proof of 1.3.

1.9 Theorem. *Let* ν *and* μ *be functions defined and measurable in* $\mathbb{C}$ *and satisfying* 1.2. *There exist two functions* $\phi_1(t, z)$, $\phi_2(t, z)$, *defined a.e. in* $\mathbb{C} \times \mathbb{C}$ *and having the following properties:*

(I) *Being functions of the variable* t, $\phi_1(t, z)$ *and* $\phi_2(t, z)$ *belong to* L_q *for every fixed* $z \in \mathbb{C}$ *and for every* $q \in [1, 2)$; *also* $\operatorname{supp}(|\phi_1(\cdot, z)| + |\phi_2(\cdot, z)|) \subseteq \mathfrak{R}$.

(II) *For every* $g \in B_p(D)$, *where* $2 < p \leqq 2 + \delta$ *and* $D \supset \mathfrak{K}$, *it holds*

$$Ug(z) := g(z) + \frac{1}{\pi} \iint \{\phi_1(t, z)\, g_t + \phi_2(t, z)\, \overline{g_t}\}\, d\sigma_t = V^{-1}g(z)$$

(here we again put $\phi_1(t, z) f(t) = \phi_2(t, z) f(t) = 0$ $\forall t \notin \mathfrak{K}$ *and every function* f; V^{-1} *means the inverse operator of* V; *let us recall that an integral with no domain of integration specified is to be considered as an integral over the whole support of the integrand).*

(III) *If* $g \in B_p(D)$ $(p, D$ *as in* (II)) *is analytic in an open subset* $D' \subseteqq D$, *then* Ug *is a* (ν, μ)-*solution in* D'; *in particular,* U *is a one-to-one mapping of* $A(D)$ *onto* $\mathfrak{A}(D, \nu, \mu)$.

(IV) $\phi_1(\cdot, z), \phi_2(\cdot, z)$ *with the properties stated in* (I), (II), *are uniquely determined as elements of* L_q, *for arbitrary but fixed* $z \in \mathbb{C}$.

Proof. Let $L_{q,0}(D) = \{f \colon f \in L_q(D), \operatorname{supp} f \Subset D\}$.

1.10 *If for any fixed functions* $h_1, h_2 \in L_{q,0}(D)$, $1 < q < 2$, *and for every* $f \in B_p(D)$, $p^{-1} + q^{-1} = 1$, *it holds*

$$\iint [h_1(z) f_z(z) + h_2(z) \overline{f_z(z)}]\, d\sigma_z = 0,$$

then $h_1(z) = h_2(z) = 0$ *a.e. in* D.

Together with f, $i \cdot f$ also belongs to $B_p(D)$, hence $\iint h_1 f_z\, d\sigma_z = 0$ $\forall f \in B_p(D)$. Now let g be any function from $L_p(D)$. There exists an $f \in B_p(D)$ such that $f_z = g$, namely $(\overline{P_{0D}\bar{g}})$ (see 0.4.19, 0.4.29). Thus we have $\iint h_1 g\, d\sigma = 0$ $\forall g \in L_p(D)$, which means that $h_1(z) = 0$ a.e. in D. In an analogous way we conclude that $h_2(z) = 0$ a.e. in D.

Now let ϕ_1, ϕ_2 be two functions in $\mathbb{C} \times \mathbb{C}$, satisfying (I) and (II). Then $\phi_1(\cdot, z), \phi_2(\cdot, z)$ are uniquely determined as elements of $L_{q,0}(\mathbb{C})$ for arbitrary fixed $z \in \mathbb{C}$. Suppose, there were two pairs ϕ_1, ϕ_2 and ϕ_1', ϕ_2' having the stated properties. By (II) we should have

$$\iint [(\phi_1(t, z) - \phi_1'(t, z))\, g_t + (\phi_2(t, z) - \phi_2'(t, z))\, \overline{g_t}]\, d\sigma_t = 0 \quad \forall z \in \mathbb{C},\ \forall g \in B_p(\mathbb{C}).$$

From 1.10 we then obtain the assertion (IV).

Further, let D be an arbitrary but fixed domain satisfying

1.11 $\mathfrak{K} \Subset D \subseteqq \mathbb{C}$, let p be any fixed number $\in (2, 2 + \delta]$.

By 1.3, U is the inverse operator of V if and only if U is a left-inverse operator of V, i.e. if $UVg = g$ $\forall g \in B_p(D)$. If U is a left-inverse operator of V, then by 1.3(I) and 0.4.44 it must hold

1.12
$$0 = \iint \Big\{ f_t \left[\frac{\nu(t)}{t - z} + \phi_1(t, z) - \nu(t)\, T\phi_1(\cdot, z)(t) - \overline{\mu(t)}\, \overline{T}\phi_2(\cdot, z)(t) \right] + \overline{f_t} \left[\frac{\mu(t)}{t - z} + \phi_2(t, z) - \mu(t)\, T\phi_1(\cdot, z)(t) - \overline{\nu(t)}\, \overline{T}\phi_2(\cdot, z)(t) \right] \Big\}\, d\sigma_t,$$

where $\bar{T}h := \overline{T\bar{h}}$. Conversely, if ϕ_1, ϕ_2 have the properties mentioned in (I) and if 1.12 holds $\forall f \in B_p(D)$, then UV is the identical mapping on $B_p(D)$, that is $U = V^{-1}$. Therefore we have to show that there exist ϕ_1, ϕ_2 possessing the properties mentioned under (I) and satisfying 1.12 $\forall f \in B_p(D)$. However by 1.10 this is possible only if

$$\textbf{1.13} \quad \phi_1(t, z) = -\frac{\nu(t)}{t - z} + \nu(t)\, T\phi_1(\cdot, z)\,(t) + \overline{\mu(t)}\, \bar{T}\phi_2(\cdot, z)\,(t),$$

$$\phi_2(t, z) = -\frac{\mu(t)}{t - z} + \mu(t)\, T\phi_1(\cdot, z)\,(t) + \overline{\nu(t)}\, \bar{T}\phi_2(\cdot, z)\,(t).$$

Conversely, if $\phi_1(\cdot, z), \phi_2(\cdot, z)$ are L_q-solutions of 1.13 for each fixed $z \in \mathbb{C}$ and for every $q \in [2 - \delta, 2)$, then ϕ_1, ϕ_2 have the properties required in (I), and 1.12 holds $\forall f \in B_p(D)$.

The system 1.13 can be solved immediately by the Banach fixed point theorem. In view of further properties of ϕ_1, ϕ_2 it is, however, more convenient to carry the system 1.13, by introducing the new unknowns

$$\textbf{1.14} \quad F = \phi_1 + \overline{\phi_2}, \qquad G = \phi_1 - \overline{\phi_2},$$

into the equivalent system consisting of the two separated equations

$$\textbf{1.15} \quad F(t, z) = -\left(\frac{\nu(t)}{t - z} + \overline{\frac{\mu(t)}{t - z}}\right) + \nu(t)\, TF(\cdot, z)\,(t) + \mu(t)\, \overline{TF(\cdot, z)\,(t)},$$

$$\textbf{1.16} \quad G(t, z) = -\left(\frac{\nu(t)}{t - z} - \overline{\frac{\mu(t)}{t - z}}\right) + \nu(t)\, TG(\cdot, z)\,(t) - \mu(t)\, \overline{TG(\cdot, z)\,(t)}.$$

Because of

$$\textbf{1.17} \quad \|\nu Tf \pm \bar{\mu}\overline{Tf}\|_{L_q} \leqq kC(q)\, \|f\|_{L_q} \text{ and } \frac{1}{t - z} \in L_q(\mathfrak{K}) \text{ for every } q \in (1, 2)$$

the equations 1.15 and 1.16 have a unique solution F and G, respectively, for every $q \in [2 - \delta, 2)$, by the Banach fixed point theorem, and

$$\textbf{1.18} \quad \operatorname{supp}\,(|F(\cdot, z)| + |G(\cdot, z)|) \subseteqq \mathfrak{K} \quad \text{for every } z \in \mathbb{C}\,.$$

(1.18 is an immediate consequence of 1.15, 1.16). Because of the boundedness of $\mathfrak{K}$ we then have $F(\cdot, z), G(\cdot, z) \in L_q$ $\forall q \in [1, 2)$. From F and G we obtain ϕ_1, ϕ_2 with the properties required. (I) and (II) are proved.

Assertion (III) now at once follows from $VUg = g$ $\forall g \in B_p(D)$, in conjunction with the definition of V and 0.4.19(I). Theorem 1.9 is proved.

By means of the operator U, solutions of inhomogeneous systems,

$$\textbf{1.19} \quad f_{\bar{z}} = \nu f_z + \mu \overline{f_z} + c(z),$$

can also be represented easily. Namely, in consequence of theorem 1.9, we obtain

1.20 Theorem. *Let the functions* ν, μ *measurable in* $\mathbb{C}$ *satisfy* 1.2, *let D satisfy* 1.11, *and let* $c(z) \in L_p(D)$, $p > 2$. *Then*

$$\{U(g + P_{0D}c) : g \in A(D)\}$$

is the set of all solutions of 1.19 *in* D.

Proof. Observe first that $C^*(z) := P_{0D}c(z) \in B_p(D) \subseteqq B_{p'}(D)$ if $2 < p' \leqq p$. Also, by the representation theorem II.2.1, every solution f of 1.19 in D is an $L_{p'}$-solution of 1.19 in D with $p' \in (2, 2 + \delta]$ (applying the representation theorem II.2.1 to 1.19 in bounded subdomains $D' \subset D$), hence f belongs to $B_{p'}(D)$. Because of $(Vf)_{\bar{z}} = f_{\bar{z}} - \nu f_z - \mu \overline{f_z} = c$ and by 0.3.13, 0.4.19 we obtain $Vf = P_{0D}c + g$ with $g \in A(D)$. Thus $f = U(g + C^*)$, by 1.9(II). Conversely, in the same way we see that $U(g + C^*)$ is always a solution of 1.19 in D if $g \in A(D)$. This proves theorem 1.20.

1.21 *Let* $f \in L_p(D)$, $p > 2$, $D \supseteq \mathfrak{K}$. *Then, for* $j = 1, 2$,

$$\iint \phi_j(t, z) f(t)\, d\sigma_t$$

being a function of z is analytic in $\mathbb{C} \setminus \mathfrak{K}$ *and has a zero at* ∞.

Proof. As in the proof of 1.10 we find a $g \in B_p(D)$ such that $g_{\bar{z}} = f$. By 1.3(I), 1.9(II), for $h := Ug \in B_{p'}(D)$, $2 < p' \leqq p$, we have

$$\textbf{1.22} \quad \frac{1}{\pi}\iint (\phi_1(t, z) f(t) + \phi_2(t, z)\overline{f(t)})\, d\sigma_t = -\frac{1}{\pi}\iint \frac{\nu(t)\, h_t + \mu(t)\, \overline{h_t}}{t - z}\, d\sigma_t.$$

From this, and by considering $i \cdot f$ and $\bar{f}$ in place of f, the assertions in 1.21 follow.

IV.2 SOME SPECIAL CASES

Concerning schlicht (ν, μ)-solutions, with the aid of the integral operators just constructed we can, for instance, obtain the following.

2.1 Theorem. *Let A, B be arbitrary constants from* $\mathbb{C}$, $A \neq 0$, *also,* ν, μ *are to satisfy the conditions* 1.2. *Then*

$$\text{(I)}\ f(z) = Az + B + \frac{1}{\pi}\iint (\phi_1(t, z)\, A + \phi_2(t, z)\, \bar{A})\, d\sigma_t$$

is that unique (ν, μ)*-solution in* $\mathbb{C}$ *which has a Laurent expansion*

$$\text{(II)}\ f(z) = Az + B + \frac{a_1}{z} + \frac{a_2}{z^2} + \dots$$

at ∞. *This* $f(z)$ *is schlicht in* $\mathbb{C}$.

Proof. By theorem 1.9 $f(z)$ is a (ν, μ)-solution in $\mathbb{C}$, and by 1.21 it has a Laurent expansion 2.1(II) at ∞. Suppose there were two (ν, μ)-solutions in $\mathbb{C}$ having a Laurent expansion 2.1(II) (with the same A and B). Their difference then would also be a (ν, μ)-solution in $\mathbb{C}$ and would have a zero at ∞, hence it should be identical to zero. The schlichtness of f in $\mathbb{C}$ follows at once from the representation $f = F \circ \chi$ where χ is a quasiconformal mapping of $\mathbb{C}$ onto itself and F is analytic in $\mathbb{C}$. Indeed, by 2.1 (II) and (for example) the argument principle, ∞ must be a simple pole of F. Hence F is an entire linear transformation. This proves 2.1.

Let us now consider some special (ν, μ)-systems.

a) If $\mu \equiv 0$ then the system 1.13 reduces to

2.2 $$\phi_1(t, z) = -\frac{\nu(t)}{t - z} + \nu(t)\, T\phi_1(\cdot, z)\,(t), \qquad \phi_2(t, z) \equiv 0,$$

and further, in terms of F and G from 1.14, we then have

2.3 $$F = G = \phi_1(t, z).$$

b) If $\nu \equiv 0$ then 1.13 reads

2.4 $$\phi_1(t, z) = \overline{\mu(t)}\, \overline{T}\phi_2(\cdot, z)\,(t),$$
$$\phi_2(t, z) = -\frac{\mu(t)}{t - z} + \mu(t)\, T(\bar{\mu}\overline{T}\phi_2(\cdot, z))\,(t)$$

that is, 1.13 essentially reduces to a single equation (for ϕ_2), as in the case of a Beltrami system.

c) Similar things also hold for (ν, μ)-systems satisfying

2.5 $$|\nu(z)| = |\mu(z)| \quad \forall z \in \mathbb{C}$$

(otherwise ν, μ as in 1.2). Geometrically, 2.5 means that the S- and S^*-ellipses associated with the (ν, μ)-system according to II.3.10, have the same ratio of the main axes, cf. II.3.16. On account of 2.5 there exists a real-valued function $\varphi(z)$ such that

2.6 $$\mu(z) = \nu(z)\, \mathrm{e}^{\mathrm{i}\varphi(z)},$$

which means we may also represent ν, μ in the form

2.7 $$\nu(z) = \varkappa_1(z) \cdot \varkappa_2(z), \qquad \mu(z) = \varkappa_1(z) \cdot \overline{\varkappa_2(z)}, \quad \text{with } \varkappa_2(z) = \mathrm{e}^{-\mathrm{i}\varphi(z)/2}.$$

From 1.13 we then at once infer that

2.8 $$\phi_1(t, z)/\varkappa_2(t) = \phi_2(t, z)/\overline{\varkappa_2(t)} =: \phi_0(t, z)$$

is a solution of the singular integral equation

2.9 $$\phi_0(t, z) = -\frac{\varkappa_1(t)}{t - z} - \frac{1}{\pi}\iint \phi_0(s, z) \cdot 2\mathrm{Re}\,\frac{\varkappa_1(t) \cdot \varkappa_2(s)}{(t - s)^2}\, \mathrm{d}\sigma_s.$$

Thus the operator U takes the shape

2.10 $$Ug(z) = g(z) + \frac{1}{\pi}\iint \phi_0(t, z) \cdot 2\mathrm{Re}\,(\varkappa_2(t)\, g_t(t))\, \mathrm{d}\sigma_t.$$

Special cases of 2.5 will play a crucial role in chapter VI.

IV.3 CONSTRUCTION OF FUNDAMENTAL SOLUTIONS BY MEANS OF THE KERNELS ϕ_1, ϕ_2

By means of the solutions $F, G \in L_q$ of the equations 1.15, 1.16 we define the functions

3.1
$$M(t, z) = \log(t - z) - PF(\cdot, z)(t)$$
$$N(t, z) = \log(t - z) - PG(\cdot, z)(t).$$

Because of 1.15, 1.16 and 0.4.31 (II), in $\mathbb{C} \setminus \{z\}$ it then holds

3.2
$$M_{\bar{t}} = \nu(t)\, M_t + \overline{\mu(t)}\, \overline{M_t}, \qquad N_{\bar{t}} = \nu(t)\, N_t - \overline{\mu(t)}\, \overline{N_t}.$$

By lemma I.1.7 (II) (applied to M, N in simply-connected subdomains of $\mathbb{C} \setminus \{z\}$) in conjunction with 0.4.31 (I) it follows that, after possible alteration and extension on null sets, $M(t, z)$ and $N(t, z)$, being functions of t, are solutions of the $(\nu, \bar{\mu})$- and the $(\nu, -\bar{\mu})$-systems, respectively, in (simply-connected subdomains of) $\mathbb{C} \setminus \{z\}$. In particular, $M(t, z)$, $N(t, z)$ are (equivalent to) continuous functions of t.

3.3 ***Definition.*** Let the coefficients ν, μ satisfy 1.2. $H(t, z)$ is called the fundamental solution of the (ν, μ)-system if it holds:

(I) For arbitrary but fixed $z \in \mathbb{C}$
$$r_\infty(t, z) := H(t, z) - \log(t - z)$$
is (that is, can be defined to be) a single-valued continuous function of t in $\mathbb{C} \setminus \{z\}$,

(II) for every z as in (I) $r_\infty(t, z)$, as a function of t, belongs to $L_r(K)$ for every finite $r \geqq 1$ and every set K bounded and measurable,

(III) for every z as in (I) $H(t, z)$ is a (ν, μ)-solution in (every simply-connected subdomain of) $\mathbb{C} \setminus \{z\}$,

(IV) for every z as in (I)
$$\lim_{t \to \infty} r_\infty(t, z) = 0.$$

3.4 *$H(t, z)$ is uniquely determined by the properties* 3.3 (I)–(IV).

Proof. Suppose, there are two functions $H_1(t, z)$, $H_2(t, z)$ with these properties. Then $H_0(t, z) := H_1 - H_2$ is a (single-valued) (ν, μ)-solution in $\mathbb{C} \setminus \{z\}$ for arbitrary but fixed z, by (I) and (III). By the representation theorem $H_0 = f \circ \chi(t)$, where χ is a quasiconformal mapping of $\mathbb{C}$ onto itself and f is analytic in $\mathbb{C} \setminus \{\chi(z)\}$. At the point $\chi(z)$, f cannot have an essential singularity, by (II) and lemma I.3.22 ((II) and I.3.22 imply a restricted growth of $H_0(t, z)$ as $t \to z$; because of the Hölder continuity of $\chi(t)$ this gives a restricted growth of $f(\chi)$ as $\chi \to \chi(z)$). Again by (II) and the Hölder continuity of $\chi(t)$, $\chi(z)$ cannot be a pole, either. Hence $\chi(z)$ is a removable singularity of f. By (IV) and Liouville's theorem f, and therefore also $H_0(t, z)$, must be identical to zero, which we had to prove.

Until further notice, the coefficients ν, μ are to satisfy 1.2.

3.5 Theorem. *$M(t, z)$ and $N(t, z)$ from 3.1 are the fundamental solutions of the $(\nu, \bar{\mu})$- and $(\nu, -\bar{\mu})$-systems, respectively. Between the kernels $\phi_1(t, z)$, $\phi_2(t, z)$ of the (ν, μ)-system and the fundamental solutions M and N of the adjoint and $(-)$-adjoint systems there exist the relations*

$$\text{(I)}\ \phi_1(t, z) = -\tfrac{1}{2}(M + N)_{\bar{t}},\ \phi_2(t, z) = -\tfrac{1}{2}\overline{(M - N)}_t$$

as well as

$$\text{(II)}\ M(t, z) = \log(t - z) - P(\phi_1(\cdot, z) + \overline{\phi_2(\cdot, z)})(t),$$

$$N(t, z) = \log(t - z) - P(\phi_1(\cdot, z) - \overline{\phi_2(\cdot, z)})(t).$$

Proof. The validity of 3.5(I), (II) is clear because of 1.14 and 0.4.31(II). In order to prove that M and N are fundamental solutions, it still remains to show (observe the considerations following 3.2) that 3.3(II), 3.3(IV) are satisfied. But this is a consequence of $F, G \in L_q$ $\forall q \in [1, 2)$, in conjunction with 0.4.31 and 1.18.

In addition to 1.2, on ν, μ we now put the condition

3.6 $$\nu, \mu \in HL_p(\{z_0\}),\quad 2 < p < \infty.$$

Under this additional condition we want to characterize the behaviour of $M(t, z_0)$, $N(t, z_0)$ as $t \to z_0$ (of course, only $z_0 \in \mathfrak{K}$ is interesting).

By means of the affine transformations

3.7 $$s = t + \ell\overline{(t - z_0)},\qquad h = H + \bar{b}\bar{H}$$

with ℓ, b as in II.3.38, the $(\nu, \bar{\mu})$-system for M goes into a $(\nu_2, \bar{\mu}_2)$-system for $m(s, z_0) := M(t(s), z_0) + \bar{b}\overline{M(t(s), z_0)}$, where

3.8 $$\nu_2(s) = |\nu(t(s)) - \nu(z_0)| \cdot B_1(s, z_0) + |\mu(t(s)) - \mu(z_0)| \cdot B_2(s, z_0)$$

$$\mu_2(s) = |\nu(t(s)) - \nu(z_0)| \cdot B_3(s, z_0) + |\mu(t(s)) - \mu(z_0)| \cdot B_4(s, z_0),$$

$B_j(s, z_0)$ is bounded and measurable, $j = 1$ to 4, cf. II.3.37. Hence

3.9 $$\nu_2, \mu_2 \in HL_p(\{z_0\})\quad\text{and}\quad \nu_2(z_0) = \mu_2(z_0) = 0$$

(however, ν_2 and μ_2 no longer have bounded support). Let α be an arbitrary constant. By the Banach fixed point theorem (note $|\nu_2(s)| + |\mu_2(s)| \leqq k' < 1$, $k' = 1 - \dfrac{(1 - k)^3}{(1 + k)^2}$, cf. II.3.37, III.4.24) the equation

3.10

$$F_\alpha(s, z) = -\left(\frac{\alpha\nu_2(s)}{s - z_0} + \overline{\frac{\alpha\mu_2(s)}{s - z_0}}\right) + \nu_2(s)\, TF_\alpha(\cdot, z_0)(s) + \overline{\mu_2(s)}\,\overline{TF_\alpha(\cdot, z_0)(s)}$$

has exactly one solution in $L_{p'}$ with certain $p' \in (2, p]$, and

3.11 $$M_\alpha(s, z_0) = \alpha \log(s - z_0) - P_0F_\alpha(\cdot, z_0)(s)$$

then is a $(\nu_2, \bar{\mu}_2)$-solution in $\mathbb{C} \setminus \{z_0\}$. Consequently,

3.12 $$M_*(t, z_0) := M_\alpha(t + \ell\overline{(t - z_0)}, z_0) - \bar{b}\overline{M_\alpha(t + \ell\overline{(t - z_0)}, z_0)}$$

is a $(\nu, \bar{\mu})$-solution in $\mathbb{C} \setminus \{z_0\}$, and we have

3.13 $$M_*(t, z_0) = \alpha \log (t - z_0 + \ell(\overline{t - z_0})) - \bar{b}\bar{\alpha} \log (\overline{t - z_0 + \bar{\ell}(t - z_0)}) \\ - P_0F_\alpha(\cdot, z_0)(t + \ell(\overline{t - z_0})) + \bar{b}\overline{P_0F_\alpha(\cdot, z_0)(t + \ell(\overline{t - z_0}))}.$$

Now let α satisfy

3.14 $$\alpha + \bar{b}\bar{\alpha} = 1, \quad \text{i.e. } \alpha = (1 - \bar{b})/(1 - |b|^2).$$

Then,

3.15 $$S(t, z_0) := M(t, z_0) - M_*(t, z_0)$$

(with $M(t, z_0)$ from 3.1) is a single-valued solution of the $(\nu, \bar{\mu})$-system in $\mathbb{C} \setminus \{z_0\}$, belonging to $L_{r,\mathrm{loc}}$ for every $r \geqq 1$. As in the proof of 3.4 we infer that $t = z_0$ is a removable singularity of $S(t, z_0)$. From 3.5, 3.3(IV), 0.4.3(II) we further obtain

3.16 $$|S(t, z_0)| \leqq \text{constant} \cdot |t|^\gamma, \quad \text{e.g. for } |t| > 1,\ \gamma \text{ arbitrary } > 0.$$

Hence, by theorem III.2.7, $S(t, z_0)$ is constant in $\mathbb{C}$. This implies

3.17 $$M(t, z_0) = \alpha \log (t - z_0 + \ell(\overline{t - z_0})) - \bar{\alpha}\bar{b} \log (\overline{t - z_0 + \bar{\ell}(t - z_0)}) \\ + (\text{Hölder continuous function of } t \text{ in } \mathbb{C}),$$

with α from 3.14 and ℓ, b from II.3.38. In a corresponding way we see that

3.18 $$N(t, z_0) = \alpha_N \log (t - z_0 + \ell(\overline{t - z_0})) + \overline{\alpha_N}\bar{b} \log (\overline{t - z_0 + \bar{\ell}(t - z_0)}) \\ + (\text{Hölder continuous function of } t \text{ in } \mathbb{C})$$

with $\alpha_N = (1 + \bar{b})/(1 - |b|^2)$ and ℓ, b from II.3.38 (one has only to replace μ by $-\mu$ and, therefore, b by $-b$ in the above considerations which led to 3.17).

IV.4 FUNDAMENTAL SOLUTIONS IN MORE GENERAL CASES

Let the function ν, μ be defined and measurable in all $\mathbb{C}$ (which is no loss in generality), and let

4.1 $$|\nu(z)| + |\mu(z)| \leqq \text{constant} =: k < 1 \quad \forall z \in \overline{\mathbb{C}}.$$

For such ν, μ the equations 1.15 and 1.16 have unique solutions $F(t, z)$, $G(t, z)$ in $L_{p,q}$ for every p, q satisfying

4.2 $$2 - \delta \leqq q < 2 < p \leqq 2 + \delta, \quad \delta \text{ as in III.4.5, for instance.}$$

This means that there are $F_1(t, z)$, $G_1(t, z) \in L_q$, $F_2(t, z)$, $G_2(t, z) \in L_p$ so that

4.3 $$F(t, z) = F_1(t, z) + F_2(t, z), \qquad G(t, z) = G_1(t, z) + G_2(t, z).$$

We set

4.4 $$M^*(t, z) = \log (t - z) - P^*(F_1(\cdot, z) + F_2(\cdot, z))(t), \\ N^*(t, z) = \log (t - z) - P^*(G_1(\cdot, z) + G_2(\cdot, z))(t)$$

(cf. 0.4.39). On account of 0.4.42 M^* and N^*, being functions of t in $\mathbb{C} \setminus \{z\}$, possess L_q-derivatives, and by 1.15 and 1.16 we have

4.5 $$M_{\bar{t}}^* = \nu(t)\, M_t^* + \overline{\mu(t)}\, \overline{M_t^*}, \qquad N_{\bar{t}}^* = \nu(t)\, N_t^* - \overline{\mu(t)}\, \overline{N_t^*}.$$

Again from lemma I.1.7(II) (while using 0.4.39 and 0.4.31 as well as 0.4.3(III)) we conclude that (after possible alteration and extension on null sets) M^* and N^* are $(\nu, \bar{\mu})$- and $(\nu, -\bar{\mu})$-solutions, respectively, in $\mathbb{C} \setminus \{z\}$, and that, in particular, they are continuous there. $M^*(t, z)$ and $N^*(t, z)$ satisfy 3.3(I)–(III) of the definition of the fundamental solution of a $(\nu, \bar{\mu})$- and a $(\nu, -\bar{\mu})$-system, respectively. Moreover, the expressions for M^* and N^* in 4.4 differ only by constants (for z fixed) if we use different decompositions 4.3 of $F(t, z)$ and $G(t, z)$, cf. 0.4.40. This means that, in a sense, each $M^*(t, z)$ and each $N^*(t, z)$ can be seen as a fundamental solution of the $(\nu, \bar{\mu})$- and the $(\nu, -\bar{\mu})$-system, respectively, if the only condition required for ν, μ is 4.1.

Next we want to consider fundamental solutions under condition 4.1 and, additionally,

4.6 $$\nu, \mu \in HL_p(\{\infty\}), \quad p > 2.$$

For instance, 4.6 is satisfied if $\nu - \nu(\infty)$, $\mu - \mu(\infty) \in L_{p^*}$ with a p^* satisfying $p^* > p$, $\dfrac{2}{p} + \dfrac{2}{p^*} > 1$.

By means of the affine mappings

4.7 $$s = t + \ell(\overline{t - z}), \qquad h = H + b\bar{H}$$

the (ν, μ)-system $H_{\bar{t}} = \nu H_t + \mu \overline{H_t}$ is carried into a system $h_{\bar{s}} = \nu_2 h_s + \mu_2 \overline{h_s}$; here b and ℓ are to be chosen as in II.3.38, where we only have to replace z_0 by ∞. Then ν_2 and μ_2 are of the shape 3.8, again with z_0 replaced by ∞. In particular $\nu_2(\infty) = \mu_2(\infty) = 0$.

Now we put

4.8 $$w = \frac{1}{s - z}, \qquad \nu_*(w) = \nu_2\left(\frac{1}{w} + z\right), \qquad \mu_*(w) = \mu_2\left(\frac{1}{w} + z\right).$$

An easy computation shows that $\nu, \mu \in HL_p(\{\infty\})$ implies $\nu_*, \mu_* \in HL_p(\{0\})$ for every fixed $z \in \mathbb{C}$. Under the inversion 4.8 the (ν_2, μ_2)-system goes into

4.9 $$H_{\bar{w}}^* = \nu^* H_w^* + \mu^* \overline{H_w^*} \quad \text{where} \quad \nu^*(w) = \nu_*(w)\left(\frac{w^2}{\bar{w}^2}\right),$$

$$\mu^*(w) = \mu_*(w), \quad H^*(w) = h\left(\frac{1}{w} + z\right);$$

since $\nu_*(0) = 0$, also $\nu^*, \mu^* \in HL_p(\{0\})$.

We now fix any constant α from $\mathbb{C}$. By the Banach fixed point theorem the equation

4.10 $$F_\alpha(w) = \left(\frac{\alpha \nu^*(w)}{w} + \frac{\bar{\alpha}\mu^*(w)}{\bar{w}}\right) + \nu^*(w)\, TF_\alpha(\cdot)\,(w) + \mu^*(w)\, \overline{TF_\alpha(\cdot)\,(w)}$$

again has just one solution F_α in $L_{p'}$ with a $p' \in (2, p]$, and

4.11 $$H_\alpha(w) := -\alpha \log w - P_0 F_\alpha(w)$$

then is a (ν^*, μ^*)-solution in $\mathbb{C} \setminus \{0\}$. Hence

4.12 $$\tilde{H}_\alpha(s, z) = H_\alpha\left(\frac{1}{s - z}\right)$$

is a (ν_2, μ_2)-solution in $\mathbb{C} \setminus \{z\}$, and, finally,

4.13 $$H_\alpha(t, z) = \tilde{H}_\alpha(t + \ell(\overline{t - z}), z) - b\overline{\tilde{H}_\alpha(t + \ell(\overline{t - z}), z)}$$

is a (ν, μ)-solution in $\mathbb{C} \setminus \{z\}$, where

4.14 $$H_\alpha(t, z) = \alpha \log (t - z + \ell(\overline{t - z})) - b\bar{\alpha} \log (\overline{t - z} + \bar{\ell}(t - z))$$
$$- P_0 F_\alpha\left(\frac{1}{t - z + \ell(\overline{t - z})}\right) + b\overline{P_0 F_\alpha\left(\frac{1}{t - z + \ell(\overline{t - z})}\right)}.$$

As to the constant α, analogously with 3.14 we put

4.15 $\alpha := \alpha_\infty := (1 - b_\infty)/(1 - |b_\infty|^2)$, $b_\infty = b$ with the same b as in 4.7 above. Let $H_*(t, z) := H_{\alpha_\infty}(t, z)$.

4.16 *$H_*(t, z)$ is the fundamental solution of the (ν, μ)-system under the conditions* 4.1, 4.6 *in the sense of the following.*

4.17 ***Definition.*** Let the coefficients ν, μ satisfy 4.1, 4.6. $H(t, z)$ is called the fundamental solution of the (ν, μ)-system, if

(I) for arbitrary but fixed $z \in \mathbb{C}$

$$r_\infty(t, z) := H(t, z) - S_\infty(t, z, \nu, \mu)$$

being a function of t is single-valued and continuous in $\mathbb{C} \setminus \{z\}$, where

$$S_\infty(t, z, \nu, \mu) := \alpha_\infty \log (t - z + \ell_\infty(\overline{t - z})) - b_\infty\bar{\alpha}_\infty \log (\overline{t - z} + \bar{\ell}_\infty(t - z)),$$

$\ell_\infty := \ell$ as in 4.7, α_∞ and b_∞ as in 4.15,

(II) with this $r_\infty(t, z)$ 3.3(II)–(IV) are satisfied.

As in the proof of 3.4 we conclude that $H(t, z)$ is uniquely determined by 4.17(I), (II). We still have to prove 4.16.

Proof (of 4.16). Clearly, $r_\infty(t, z) := H_*(t, z) - S_\infty(t, z, \nu, \mu)$ satisfies 4.17(I). By 4.14, 0.4.3(II), 4.10, 4.6 we further have

4.18 $$|r_\infty(t, z)| = (1 + |b|)\, K_{p'} \|F_\alpha\|_{L_{p'}} \cdot (1 - |\ell|)^{-1}\, |t - z|^{-(1-(2/p'))}$$

where p' is arbitrary from $(2, 2 + \delta]$. Hence $r_\infty(t, z)$ also satisfies 3.3(II) and 3.3(IV). Furthermore, by construction, $H_*(t, z)$ is a (ν, μ)-solution in $\mathbb{C} \setminus \{z\}$, so that 3.3(III) also holds. Consequently, $H_*(t, z) = H(t, z)$ is the unique fundamental solution of the (ν, μ)-system (in the sense of definition 4.17).

Beyond 4.1 and 4.6, let

4.19 $$\nu, \mu \in HL_p(\{z_0\}), \quad p > 2.$$

Then $H(t, z)$ shows not only a certain asymptotic behaviour as $t \to \infty$ (for any $z \in \mathbb{C}$) but $H(t, z_0)$ also shows a well-defined asymptotic behaviour as $t \to z_0$, namely analogous with 3.17 we then have

4.20 $$H(t, z_0) = \alpha \log (t - z_0 + \ell\overline{(t - z_0)}) - \bar{\alpha} b \log (\overline{t - z_0} + \bar{\ell}(t - z_0)) + (\text{Hölder continuous function of } t \text{ in } \mathbb{C})$$

with b, ℓ from II.3.38 and $\alpha = (1 - b)/(1 - |b|^2)$. To prove this we only need to repeat the considerations in 3.6 to 3.14 (with $\bar{\mu}, \bar{b}$ replaced by μ, b, respectively). This leads to a (ν, μ)-solution $M_*(t, z_0)$ in $\mathbb{C} \setminus \{z_0\}$ with an asymptotic expansion as on the left-hand side of 4.20 (or as in 3.13 with $\bar{b}$ replaced by b). Because of 4.15 and 3.14, $H(t, z_0) - M_*(t, z_0)$ is single-valued in $\mathbb{C} \setminus \{z_0\}$. As in the proof of 3.4 or 4.16 we conclude that z_0 is a removable singularity.

By 3.13, 4.17 we have

$$|H(t, z_0) - M_*(t, z_0)| \leqq K_\gamma |t|^\gamma \quad \text{for any } \gamma > 0, \quad |t| > 1,$$

where K_γ is constant. Hence $H(t, z_0) - M_*(t, z_0) = \text{constant}$, by theorem III.2.7, which proves 4.20.

Analogously as with $S_\infty(t, z, \nu, \mu)$ we put

4.21 $$S(t, z_0, \nu, \mu) = \alpha \log (t - z_0 + \ell\overline{(t - z_0)}) - \bar{\alpha} b \log (\overline{t - z_0} + \bar{\ell}(t - z_0)),$$

with α, ℓ, b as in 4.20.

For later use we place at our disposal a lemma which states a kind of uniformity with respect to the second variable z of fundamental solutions.

4.22 Lemma. *Let D be a bounded subset of $\mathbb{C}$. If ν, μ satisfy* 4.1 *as well as $\nu, \mu \in HL_p(D \cup \{\infty\})$, $p > 2$, then the fundamental solution $H(t, z)$ of the (ν, μ)-system admits the decomposition*

(I) $H(t, z_0) = S(t, z_0, \nu, \mu) + h(t, z_0) + \mathfrak{K}(z_0)\ \forall t, z_0 \in \mathbb{C},\ t \neq z_0,$

with the following properties: $\mathfrak{K}(z_0)$ depends only on z_0, $h(t, z_0)$ is a single-valued and Hölder continuous function of t in $\mathbb{C}$ for every $z_0 \in D$, and for every bounded domain G' there exists a constant $K(G') < \infty$ so that

(II) $|h(t, z_0)| \leqq K(G')\ \forall t \in G',\ \forall z_0 \in D.$

Proof. The coefficients ν_2, μ_2 with $\nu_2(z_0) = \mu_2(z_0) = 0$ which belong to ν, μ, z_0 according to 3.7, 3.8 (with b in place of $\bar{b}$) have uniformly bounded $HL_p(\{z_0\})$-norms for every $z_0 \in D$. With this ν_2, μ_2 we solve equation 3.10 (with $\overline{\mu_2}$ replaced by μ_2). Its solutions $F_\alpha(\cdot, z_0)$ have uniformly bounded L_p-norms for all uniformly bounded α and every $z_0 \in D$ (cf., e.g. III.4.24). In an analogous way as with 3.11 to 3.16 the asserted representation (I) with the stated properties follows.

Chapter V

INTEGRAL FORMULAE FOR (ν, μ)-SOLUTIONS

In this chapter some integral theorems of the classical function theory shall be carried over to (ν, μ)-solutions. First we prove the generalized Cauchy integral formula (which was known heretofore only under stronger conditions on the coefficients ν, μ, cf. [P2] and the corresponding literature cited there as well as [S1]). This formula will then be used in order to derive relations between fundamental solutions. After this we consider generalized Cauchy type integrals (as to the special case $\nu \equiv 0$, cf. [P2], p. 82) and prove a generalization of the classical saltus conditions. In the fourth (and last) section of this chapter the Schwarz and Poisson integral formulae are generalized to (ν, μ)-solutions (concerning a similar relation in the special case $\nu \equiv 0$, cf. [K4]). As in the classical case, here a reflection process plays a role. This fact requires some additional conditions on ν, μ. Furthermore we show that (under these conditions on ν, μ) the generalized Poisson integral solves the problem of giving the real part of that (of course, unique) (ν, μ)-solution which takes prescribed piecewise continuous boundary values on the unit circle.

V.1 THE GENERALIZED CAUCHY INTEGRAL FORMULA

Let ν, μ be defined and measurable in $\overline{\mathbb{C}}$, which is no loss of generality, and let IV.4.1, IV.4.6 hold. Let $H(t, z, \nu, \mu)$ be the fundamental solution of the (ν, μ)-system according to IV.4.17. We put

1.1 $$M(t, z) := H(t, z, \nu, \bar{\mu}), \qquad N(t, z) := H(t, z, \nu, -\bar{\mu}).$$

Let G, D be as in 0.3.17, and let G be a domain as well. The coefficients ν, μ additionally are to satisfy $\nu, \mu \in HL_p(\partial D \cup (\{|z - z_0| < d\} \cap D))$, where $2 < p$, $d > 0$ and z_0 is any fixed point from $\mathbb{C}$. Then for every (ν, μ)-solution $f(z)$ in G we have the (generalized) Cauchy integral formula

1.2

$$\operatorname{Re}\frac{1}{2\pi i}\int_{\partial D} f(t)\,\mathrm{d}_t M(t, z_0) + i\operatorname{Im}\frac{1}{2\pi i}\int_{\partial D} f(t)\,\mathrm{d}_t N(t, z_0) = \begin{cases} f(z_0) & \text{if } z_0 \in D \\ 0 & \text{if } z_0 \in \mathbb{C}\setminus\bar{D}. \end{cases}$$

Proof. If z_0 is outside $\mathfrak{C}_1$ (cf. 0.3.17) then 1.2 holds on account of II.6.5, since $M(t, z_0)$ and $iN(t, z_0)$ can be defined to be single-valued $(\nu, \bar{\mu})$-solutions in a domain $G' \Supset D$. If z_0 is inside $\mathfrak{C}_j$, $j = 2, \ldots, n$, then we only have to apply this consideration to a suitable cross-cut of D, cf. [N], p. 196. Thus 1.2 holds when $z_0 \in \mathbb{C} \setminus \bar{D}$. Now let $z_0 \in D$. Choose a positive δ such that $\delta < d$ and $\{|t - z_0| \leqq \delta\} \subset D$. Then $D \setminus \{|t - z_0| \leqq \delta\} = D^*$ is a domain as in 0.3.17, and 1.2 also holds with D replaced by D^* if $z_0 \in \mathbb{C} \setminus \overline{D^*}$, as proved just now. This gives

1.3
$$\operatorname{Re}\frac{1}{2\pi i}\int_{\partial D} f(t)\,\mathrm{d}_t M(t, z_0) = \operatorname{Re}\frac{1}{2\pi i}\oint_{|t-z_0|=\delta} f(t)\,\mathrm{d}_t M(t, z_0).$$

In the same way we obtain

1.4
$$\operatorname{Im}\frac{1}{2\pi i}\int_{\partial D} f(t)\,\mathrm{d}_t N(t, z_0) = \operatorname{Im}\frac{1}{2\pi i}\oint_{|t-z_0|=\delta} f(t)\,\mathrm{d}_t N(t, z_0).$$

Because of IV.4.20 we have

1.5
$$\int_{\partial D} f(t)\,\mathrm{d}_t M(t, z_0) = \oint_{|t-z_0|=\delta} f(t)\,\mathrm{d}_t[(\alpha - \bar{\alpha}b)\log|t - z_0 + \ell(\overline{t - z_0})| + i\arg(t - z_0 + \ell(\overline{t - z_0})) + R(t, z_0)]$$

where $R(t, z_0)$ is a Hölder continuous function of t in a whole neighbourhood of z_0. Since $\nu, \mu \in HL_p(D \cap \{|z - z_0| < d\})$, $f(z)$ is continuously differentiable in $\{|z - z_0| < d\} \cap D$, by II.5.2(II). Hence

1.6
$$\lim_{\delta\to 0}\oint_{|t-z_0|=\delta} f(t)\,\mathrm{d}_t[(\alpha - \bar{\alpha}b)\log|t - z_0 + \ell(\overline{t - z_0})| + R(t, z_0)]$$

$$= -\lim_{\delta\to 0}\oint_{|t-z_0|=\delta}[(\alpha - \bar{\alpha}b)\log|t - z_0 + \ell(\overline{t - z_0})| + R(t, z_0)](f_t\,\mathrm{d}t + f_{\bar{t}}\,\mathrm{d}\bar{t}) = 0.$$

Further we have

1.7
$$\oint_{|t-z_0|=\delta} f(t)\,\mathrm{d}_t\arg(t - z_0 + \ell(\overline{t - z_0}))$$

$$= f(z_0)\cdot 2\pi + \oint_{|t-z_0|=\delta}(f(t) - f(z_0))\,\mathrm{d}_t\arg(t - z_0 + \ell(\overline{t - z_0})),$$

and the contour integral on the right-hand side is of absolute value

1.8 $$\leqq 2\pi \max\{|f(t) - f(z_0)| : |t - z_0| \leqq \delta\}.$$

Hence

1.9 $$\lim_{\delta\to 0} \oint_{|t-z_0|=\delta} f(t)\,\mathrm{d}_t M(t, z_0) = 2\pi \mathrm{i} f(z_0).$$

In the same way we obtain

1.10 $$\lim_{\delta\to 0} \oint_{|t-z_0|=\delta} f(t)\,\mathrm{d}_t N(t, z_0) = 2\pi \mathrm{i} f(z_0).$$

Together with 1.3, 1.4 this gives the assertion 1.2.

Now let $z_0 \in \partial D$ and ∂D have right-sided and left-sided tangents at z_0 forming the interior angle γ (with respect to D), $0 \leqq \gamma \leqq 2\pi$. Around z_0 we draw a circle of radius $\delta < d$. There is just one (maximal connected) component Z_δ of $D \cap \{|z - z_0| < \delta\}$ having z_0 as a boundary point. Let $K_\delta := \overline{Z_\delta} \cap \{|z - z_0| = \delta\} \neq \emptyset$, where $\emptyset$ is the empty set, K_δ is positively oriented (i.e. oriented in keeping with the positive orientation of $\{|z - z_0| = \delta\}$) and $\partial D_\delta := \partial D \setminus \overline{Z_\delta}$. If δ is sufficiently small we have

1.11 $$\operatorname{Re}\frac{1}{2\pi \mathrm{i}} \int_{\partial D_\delta} f(t)\,\mathrm{d}_t M(t, z_0) = \operatorname{Re}\frac{1}{2\pi \mathrm{i}} \int_{K_\delta} f(t)\,\mathrm{d}_t M(t, z_0),$$

in place of 1.3. From this, and in an analogous way as above, we conclude that, in place of 1.9,

1.12 $$\lim_{\delta\to 0} \int_{K_\delta} f(t)\,\mathrm{d}_t M(t, z_0) = \gamma \mathrm{i} f(z_0)$$

holds if, in addition, $\nu(z_0) = 0$ (this implies $\ell = \ell(z_0) = 0$). A corresponding relation holds instead of 1.4, 1.10. Thus, again under the additional assumption $\nu(z_0) = 0$, we finally obtain

1.13 $$\gamma f(z_0) = \operatorname{Re}\frac{1}{\mathrm{i}} \int_{\partial D} f(t)\,\mathrm{d}_t M(t, z_0) + \mathrm{i}\operatorname{Im}\frac{1}{\mathrm{i}} \int_{\partial D} f(t)\,\mathrm{d}_t N(t, z_0),$$

where the contour integrals in 1.13 have to be understood in the sense of the Cauchy principal value,

$$\int_{\partial D} f\,\mathrm{d}h = \lim_{\delta\to 0} \int_{\partial D_\delta} f\,\mathrm{d}h.$$

(A relation analogous to 1.13 can also be proved if $\nu(z_0) \neq 0$; this formula involves not only γ but also, in an elementary manner, $\ell = \ell(z_0)$ as well as the directions of the tangents of ∂D at z_0.)

We can write the left-hand side of 1.2 in a somewhat different form, namely

1.14

$$\frac{1}{2\pi \mathrm{i}}\left(\operatorname{Re}\int_{\partial D} f(t)\,\mathrm{d}_t N(t, z_0) + \mathrm{i}\operatorname{Im}\int_{\partial D} f(t)\,\mathrm{d}_t M(t, z_0)\right) = \begin{cases} f(z_0) & \text{if } z_0 \in D \\ 0 & \text{if } z_0 \in \mathbb{C} \setminus \bar{D}. \end{cases}$$

We put

1.15 $\operatorname{Re} M(t, z) = P^+(t, z), \qquad \operatorname{Im} M(t, z) = I^+(t, z),$
$\operatorname{Re} N(t, z) = P^-(t, z), \qquad \operatorname{Im} N(t, z) = I^-(t, z)$

and

1.16 $P^+(t, z) + \mathrm{i}I^-(t, z) = \tilde{M}(z, t), \qquad P^-(t, z) + \mathrm{i}I^+(t, z) = \tilde{N}(z, t),$

as well as

1.17 $\Omega_1(t, z) = \frac{1}{2}(M(t, z) + N(t, z)), \qquad \Omega_2(t, z) = -\frac{1}{2}\overline{(M(t, z) - N(t, z))}.$

Then 1.2, 1.14 become (with $f = u + \mathrm{i}v$, u and v real)

1.18
$$f(z) = \frac{1}{2\pi \mathrm{i}} \int_{\partial D} u(t)\, \mathrm{d}_t \tilde{N}(z, t) + \mathrm{i}v(t)\, \mathrm{d}_t \tilde{M}(z, t)$$
$$= \frac{1}{2\pi \mathrm{i}} \int_{\partial D} f(t)\, \mathrm{d}_t \Omega_1(t, z) + \overline{f(t)}\, \mathrm{d}_t \Omega_2(t, z) \quad \text{for } z \in D.$$

Since D is a bounded domain it is no restriction to assume G to be bounded and ν, μ to be identical to zero outside G. Hence,

1.19, *the relations* 1.2, 1.13, 1.18 *hold true especially with* $M(t, z), N(t, z)$ *from* IV.3.1 (*if we choose* ν, μ *so that* $\operatorname{supp}_0 (|\nu| + |\mu|) \subseteq G$, $D \Subset G \Subset \mathbb{C}$).

Because of $\nu, \mu \in HL_p(\partial D \cup (\{|z - z_0| < d\} \cap D))$, together with M, N and $\log(t - z)$ also $P\phi_l(\cdot, z)(t)$, $l = 1, 2$, is of bounded variation on ∂D as well as on every circle $|t - z_0| = \delta < d$ lying in D. Hence we may put IV.3.1 into 1.14 and calculate the contour integrals term by term. Using IV.3.5(II), 1.17, 1.18, this gives

1.20
$$\frac{1}{2\pi \mathrm{i}} \int_{\partial D} \frac{f(t)}{t - z}\, \mathrm{d}t + \frac{1}{2\pi \mathrm{i}} \int_{\partial D} [f(t)\, \mathrm{d}_t(-P\phi_1(\cdot, z)(t)) + \overline{f(t)}\, \mathrm{d}_t \overline{P\overline{\phi_2(\cdot, z)}(t)}]$$
$$= \begin{cases} f(z) & \text{if } z \in D \\ 0 & \text{if } z \in \mathbb{C} \setminus \bar{D}. \end{cases}$$

If we now have, additionally,

1.21 $\nu = \mu = 0$ in an open set $\ddot{O}$ with $\partial D \subset \ddot{O}$,

then $\phi_1 = \phi_2 = 0$, too, $\forall t \in \ddot{O}$. Hence, the left-hand side in 1.20 is equal to

$$\frac{1}{2\pi \mathrm{i}} \int_{\partial D} \frac{f(t)}{t - z}\, \mathrm{d}t + \frac{1}{2\pi \mathrm{i}} \int_{\partial D} f(t) \left(\frac{1}{\pi} \iint \frac{\phi_1(s, z)}{(s - t)^2}\, \mathrm{d}\sigma_s \right) \mathrm{d}t - \overline{f(t)}$$
$$\times \left(\frac{1}{\pi} \iint \frac{\phi_2(s, z)}{(\overline{s - t})^2}\, \mathrm{d}\sigma_s \right) \mathrm{d}\bar{t}.$$

By Fubini's theorem $\left(f(t(\tau)) \frac{\mathrm{d}t}{\mathrm{d}\tau} \phi_l(s, z)\, (s - t(\tau))^{-2} \in L_1(\mathbb{C} \times [0, |\mathfrak{C}_j|]) \right.$ for every boundary component $\mathfrak{C}_j$ of ∂D, where $|\mathfrak{C}_j|$ is the total length of $\mathfrak{C}_j$, and τ is the arc length parameter on $\mathfrak{C}_j$) this is equal to

1.22
$$\frac{1}{2\pi i}\int_{\partial D}\frac{f(t)}{t-z}\,dt+\frac{1}{\pi}\iint\left\{\phi_1(s,z)\frac{1}{2\pi i}\int_{\partial D}\frac{f(t)}{(t-s)^2}\,dt+\phi_2(s,z)\right.$$
$$\left.\times\overline{\frac{1}{2\pi i}\int_{\partial D}\frac{f(t)}{(t-s)^2}\,dt}\right\}d\sigma_s=U\left(\frac{1}{2\pi i}\int_{\partial D}\frac{f(t)}{t-z}\,dt\right),$$

where U is the operator defined in IV.1.9(II). Thus, under the additional condition 1.21, for every (ν, μ)-solution $f(z)$ in G we have

1.23
$$U\left(\frac{1}{2\pi i}\int_{\partial D}\frac{f(t)}{t-z}\,dt\right)=\begin{cases}f(z) & \text{if } z\in D\\ 0 & \text{if } z\in\mathbb{C}\setminus\bar{D}.\end{cases}$$

V.2 RELATIONS BETWEEN FUNDAMENTAL SOLUTIONS

Again let the coefficients ν, μ be defined everywhere in $\bar{\mathbb{C}}$. In a first step we will assume that ν, μ have bounded supports, i.e. that IV.1.2 holds. Further let $H(t, z, \nu, \mu)$ be the fundamental solution of the (ν, μ)-system, according to IV.4.17. Then we have

2.1 $H(z, t, \nu, \mu) = \operatorname{Re} H(t, z, \nu, \bar{\mu}) + i \operatorname{Im} H(t, z, \nu, -\bar{\mu}) - i\pi \quad \forall t, z \in \mathbb{C},\ t \neq z$

(with the branches of the argument function suitably chosen). The proof of 2.1 rests upon the Cauchy integral formula, which, however, in its form stated above requires further additional assumptions on ν, μ. Therefore we first prove the following convergence statement.

2.2 *Let the sequences* ν_n, μ_n *satisfy* $\operatorname{supp}(|\nu_n| + |\mu_n|) \subseteq \{|z| \leq R_0\}$, $|\nu_n(z)| + |\mu_n(z)| \leq$ *constant* $= k < 1$, $\forall z \in \mathbb{C}$ *and* $\forall n = 1, 2, \ldots$ *as well as* $\nu_n \to \nu$, $\mu_n \to \mu$ *a.e. in* $\mathbb{C}$ *as* $n \to \infty$. *Then there exists a subsequence such that* $H(t, z, \nu_n, \mu_n) \to H(t, z, \nu, \mu)$ *pointwise* $\forall z, t \in \mathbb{C}$ *with* $t \neq z$.

Proof. Let $F(t, z)$ be the solution of IV.1.15 with $\bar{\mu}$ replaced by μ there. In a corresponding way, let $F_n(t, z)$ arise from replacing ν by ν_n and $\bar{\mu}$ by μ_n in IV.1.15. Then we have $F_n(\cdot, z) \to F(\cdot, z)$ in L_q (for each fixed $z \in \mathbb{C}$) and $\operatorname{supp} F_n(\cdot, z) \cup \operatorname{supp} F(\cdot, z) \subseteq \{|t| \leq R_0\}$. For instance by 0.4.30, $PF_n(\cdot, z)(t) \to PF(\cdot, z)(t)$ in $L_q(\{|t| < R_0 + 1\})$. Hence there is a subsequence of $PF_n(\cdot, z)(t)$ converging to $PF(\cdot, z)(t)$ pointwise in $\{|t| < R_0 + 1\}$. For $|t| \geq R_0 + 1$ $PF_n(\cdot, z)(t) \to PF(\cdot, z)(t)$ follows from the Hölder inequality in conjunction with the convergence of $F_n(\cdot, z)$ to $F(\cdot, z)$ in L_q, $1 < q < 2$. Now, the $H(t, z, \nu_n, \mu_n) = \log(t - z) - PF_n(\cdot, z)(t)$ are locally uniformly bounded (ν_n, μ_n)-solutions in simply-connected subdomains of $\mathbb{C} \setminus \{z\}$. Hence, by theorem II.4.1, a subsequence converges to a (ν, μ)-solution locally uniformly in $\mathbb{C} \setminus \{z\}$, and by the reasons mentioned above this limit must be $\log(t - z) - PF(\cdot, z)(t) = H(t, z, \nu, \mu)$, which proves 2.2.

For instance by theorem 0.3.9, for ν, μ satisfying the conditions posed in IV.1.2 there exists a sequence of continuously differentiable ν_n, μ_n satisfying the conditions of 2.2. By 2.2 the relation 2.1 must hold if it holds for continuously differentiable ν, μ having bounded supports. The proof of 2.1 in the latter case runs as follows. By the Cauchy integral formula (after carrying out the limiting process $P' \to P$, $Q' \to Q$, $\delta \to 0$, see the figure), when

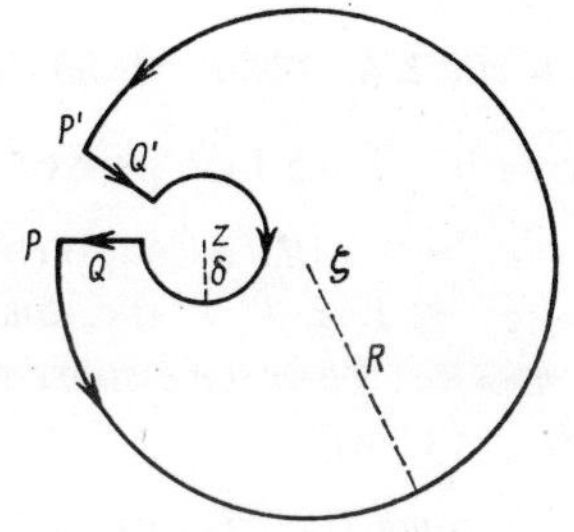

using the abbreviations $H(t, z) = H(t, z, \nu, \mu)$, $M(t, z) = H(t, z, \nu, \bar{\mu})$ we obtain

$$\textbf{2.3} \quad \operatorname{Re} H(\zeta, z) = \operatorname{Re} \frac{1}{2\pi i} \overset{\mathfrak{C}}{\int_P^P} H(t, z)\, d_t M(t, \zeta) + \operatorname{Re} M(z, \zeta) - \operatorname{Re} M(P, \zeta),$$

where $\mathfrak{C}$ is, for instance, a positively oriented circle around ζ of radius R with initial point = end point = $P = \zeta + R\,e^{i\alpha}$, containing z in its interior. Now we have (cf. section IV.3)

$$\textbf{2.4} \quad \lim_{R\to\infty} \left[\operatorname{Re} \frac{1}{2\pi i} \overset{\mathfrak{C}}{\int_P^P} H(t, z)\, d_t M(t, \zeta) - \operatorname{Re} M(P, \zeta) \right]$$

$$= \lim_{R\to\infty} \operatorname{Re} \left[\frac{1}{2\pi i} \overset{\mathfrak{C}}{\int_P^P} (\log (t - z) + r(t, z))\, d_t(\log (t - \zeta) + \tilde{r}(t, \zeta)) - M(P, \zeta) \right],$$

where $r(t, z)$ and $\tilde{r}(t, \zeta)$, with z, ζ fixed, are of the shape

$$-\frac{1}{\pi} \iint\limits_{|s| < R_0} \frac{F(s)}{s - t}\, d\sigma_s.$$

Further this is

$$= \lim_{R\to\infty} \left[\operatorname{Re} \frac{1}{2\pi i} \overset{\mathfrak{C}}{\int_P^P} \log(t - z)\, d_t \log (t - \zeta) - \operatorname{Re} M(P, \zeta) \right]$$

$$= \lim_{R\to\infty} \left[\operatorname{Re} \frac{1}{2\pi i} \int_\alpha^{\alpha+2\pi} \log (R\,e^{i\varphi} + \zeta - z)\, i\, d\varphi - \operatorname{Re} M(P, \zeta) \right]$$

$$= \lim_{R\to\infty} \left[\operatorname{Re} \frac{1}{2\pi i} \int_\alpha^{\alpha+2\pi} \log (R\,e^{i\varphi})\, i\, d\varphi - \operatorname{Re} \log (R\,e^{i\alpha}) \right] = 0;$$

the remaining terms of the integral on the right-hand side in 2.4 obviously tend to zero as $R \to \infty$. Hence $\operatorname{Re} M(z, \zeta) = \operatorname{Re} H(\zeta, z)$. In an analogous way the assertion about the imaginary parts follows. Assertion 2.1 is proved.

In a second step we now carry through the extension of 2.1 to more general ν, μ.

2.5 *Let ν, μ satisfy* IV.4.1, IV.4.6. *Then* 2.1 *holds true.*

Proof. Analogously as with 2.2 we first prove:

2.6 *Let the sequences ν_n, μ_n satisfy* $\operatorname{supp}(|\nu_n - \nu(\infty)| + |\mu_n - \mu(\infty)|) \subseteqq \{|z| < R_0\}$; $|\nu_n(z)| + |\mu_n(z)| \leqq k < 1 \ \forall z \in \mathbb{C}$ *and* $\forall n = 1, 2, \ldots$ *Further let $\nu_n \to \nu$, $\mu_n \to \mu$ a.e. in $\mathbb{C}$ as $n \to \infty$. Then for a subsequence it holds $H(t, z, \nu_n, \mu_n) \to H(t, z, \nu, \mu)$ pointwise $\forall t, z \in \mathbb{C}$ with $t \neq z$.*

For this we only have to reduce the case of ν_n, μ_n constant in a neighbourhood of ∞ by auxiliary affine mappings to the case $\nu_n = \mu_n = 0$ near ∞ and to apply 2.2. Assertion 2.6 is proved.

Next we prove 2.1 in case of, say, continuously differentiable ν, μ with $\nu =$ constant, $\mu =$ constant in a neighbourhood of ∞. To this end we again start with the relation 2.3, which of course remains valid. In place of 2.4 we have

2.7
$$\lim_{R \to \infty} \operatorname{Re}\left[\frac{1}{2\pi i} \overset{\mathfrak{C}}{\int_P^P} H(t, z)\, d_t M(t, \zeta) - M(P, \zeta)\right]$$

$$= \lim_{R \to \infty} \operatorname{Re}\left[\frac{1}{2\pi i} \overset{\mathfrak{C}}{\int_P^P} (S_\infty(t, z) + r_\infty(t, z))\, d_t(\tilde{S}_\infty(t, \zeta) + \tilde{r}_\infty(t, \zeta)) - M(P, \zeta)\right],$$

where $S_\infty(t, z) = S_\infty(t, z, \nu, \mu)$, $\tilde{S}_\infty(t, \zeta) = S_\infty(t, \zeta, \nu, \bar{\mu})$ (cf. IV.4.17) and $r_\infty(t, z) = H(t, z) - S_\infty(t, z)$, $\tilde{r}_\infty(t, \zeta) = M(t, \zeta) - \tilde{S}_\infty(t, \zeta)$. By auxiliary affine transformations of the kind IV.4.7, $r_\infty(t, z)$ and $\tilde{r}_\infty(t, z)$ change into analytic functions having zeroes at ∞. Hence the limit mentioned in 2.7 exists and is equal to

$$\lim_{R \to \infty} \left[\operatorname{Re} \frac{1}{2\pi i} \overset{\mathfrak{C}}{\int_P^P} S_\infty(t, z)\, d_t \tilde{S}_\infty(t, \zeta) - \operatorname{Re} \tilde{S}_\infty(P, \zeta)\right] = 0,$$

as an easy calculation shows. Thus we again have $\operatorname{Re} H(\zeta, z) = \operatorname{Re} M(z, \zeta)$. In the same way the assertion on the imaginary parts follows. By 2.6 and theorem 0.3.9, 2.1 then also holds for all coefficients ν, μ satisfying $|\nu(z)| + |\mu(z)| \leqq k < 1$ $\forall z \in \mathbb{C}$ and $\nu(z) =$ constant, $\mu(z) =$ constant in a neighbourhood of ∞.

In a last step to the proof of 2.5 we approximate the ν, μ mentioned in 2.5 by

$$\nu_n(z) = \nu(z), \qquad \mu_n(z) = \mu(z) \quad \text{as } |z| < n,$$
$$\nu_n(z) = \nu(\infty), \qquad \mu_n(z) = \mu(\infty) \quad \text{as } |z| \geqq n,$$

i.e. by coefficients ν_n, μ_n for which 2.1 is already known to be valid. Then from IV.4.7–16 in conjunction with 0.4.3 it follows that $H(t, z, \nu_n, \mu_n)$ converges pointwise with respect to t to $H(t, z, \nu, \mu)$ for every (fixed) $z \in \mathbb{C}$ with $t \neq z$. Of course, the same also holds for $H(t, z, \nu_n, \overline{\mu_n})$ as well as $H(t, z, \nu_n, -\overline{\mu_n})$. By changing over to the limit $n \to \infty$ in 2.1 (there with ν_n, μ_n in place of ν, μ) with fixed but arbitrary $z, t \in \mathbb{C}$, $z \neq t$, we obtain the validity of 2.1 for the limit functions, too, which is the assertion 2.5.

As a consequence, the Cauchy integral formula now also reads

2.8
$$\frac{1}{2\pi i}\int_{\partial D} u(t)\, d_t H(z, t, \nu, -\mu) + iv(t)\, d_t H(z, t, \nu, \mu) = \begin{cases} f(z) & \text{if } z \in D \\ 0 & \text{if } z \in \mathbb{C} \setminus \bar{D}. \end{cases}$$

V.3 CAUCHY INTEGRALS, SALTUS CONDITIONS

Let $\mathfrak{C}$ be a rectifiable Jordan curve or a subarc of such a Jordan curve.

3.1 *We assume that the coefficients* ν, μ *satisfy* IV.4.1, IV.4.6 *and* $\nu, \mu \in HL_p(\mathfrak{C})$ *with* $p > 2$.

Then the fundamental solutions

3.2.
$$H(t, z) := H(t, z, \nu, \mu), \qquad H^-(t, z) := H(t, z, \nu, -\mu),$$
$$M(t, z) := H(t, z, \nu, \bar{\mu}), \qquad N(t, z) := H(t, z, \nu, -\bar{\mu})$$

exist in the sense of definition IV.4.17, and all these functions are of bounded variation on $\mathfrak{C}$ with respect to t if $z \notin \mathfrak{C}$ (because of II.5.2(II) and because $\mathfrak{C}$ is rectifiable).

3.3 Theorem. *If* ν, μ *satisfy the conditions from* 3.1, *the generalized Cauchy integral*

$$\text{(I)} \quad F(z) := \frac{1}{2\pi i}\int_{\mathfrak{C}} u(t)\, d_t H^-(z, t) + iv(t)\, d_t H(z, t)$$

is a (ν, μ)*-solution in* $\mathbb{C} \setminus \mathfrak{C}$ *for any real-valued functions* $u(t), v(t) \in \mathscr{C}(\mathfrak{C})$.

Proof. The existence of the integral for each $z \in \mathbb{C} \setminus \mathfrak{C}$ at once follows from the continuity of u and v and from the (piecewise) absolute continuity of $H^-(z, t)$, $H(z, t)$ on $\mathfrak{C}$ (when using 2.1 and II.5.2(II)). We still have to prove that $F(z)$ is a (ν, μ)-solution in $\mathbb{C} \setminus \mathfrak{C}$. To this end let $t_1, \dots, t_n$ be consecutive points $\in \mathfrak{C}$ (according to the orientation of $\mathfrak{C}$). Then

3.4
$$h^{(n)}(z) := \sum_{j=1}^{n-1} v(t_j)\,[H(z, t_{j+1}) - H(z, t_j)]$$
$$= \sum_{j=1}^{n-1} v(t_j)\,[\operatorname{Re}(M(t_{j+1}, z) - M(t_j, z)) + i\operatorname{Im}(N(t_{j+1}, z) - N(t_j, z))]$$

is a (ν, μ)-solution in $\mathbb{C} \setminus \mathfrak{C}$, and with suitable choice of the t_j, $j = 1, \dots, n$, $h^{(n)}(z)$ tends to $h(z) := \int_{\mathfrak{C}} v(t)\, d_t H(z, t)$ as $n \to \infty$. The $h^{(n)}(z)$ are locally uniformly bounded in $\mathbb{C} \setminus \mathfrak{C}$. For let $t(s)$ be a parameter representation of $\mathfrak{C}$ with s the arc length parameter, and let $t_j = t(s_j)$. $M(t(s), z)$ is absolutely continuous with respect to s, hence we have

$$\textbf{3.5} \quad \operatorname{Re}(M(t_{j+1}, z) - M(t_j, z)) = \operatorname{Re} \int_{s_j}^{s_{j+1}} \frac{\mathrm{d}M(t(s), z)}{\mathrm{d}s} \mathrm{d}s$$

$$= \operatorname{Re} \int_{s_j}^{s_{j+1}} \left[M_t(t, z) \frac{\mathrm{d}t}{\mathrm{d}s} + M_{\bar{t}}(t, z) \frac{\mathrm{d}\bar{t}}{\mathrm{d}s} \right] \mathrm{d}s.$$

Again by theorem II.5.2(II), $M_t(t, z)$ and $M_{\bar{t}}(t, z)$ are uniformly bounded on $\mathfrak{C}$ (as functions of t) for every z from an arbitrary but fixed subdomain $E \Subset \mathbb{C} \setminus \mathfrak{C}$ [the $\mathfrak{P}$ from theorem II.5.2 here consists of the (single) pair $(\nu, \bar{\mu})$, and $\mathfrak{M}$ is constituted by the functions $M(t, z)$ of t with (arbitrary but fixed) $z \in E$, $\mathfrak{C}$ is subdivided (independently of the t_j) into a finite number of subarcs so that $M(t, z)$ is single-valued there, and these subarcs then act as $D = D'$ in applying theorem II.5.2; the locally uniform boundedness of the $M(t, z)$ themselves follows, for example, from IV.4.16, IV.4.18 (with μ replaced by $\bar{\mu}$)]. Hence $\operatorname{Re} h^{(n)}(z)$ is locally uniformly bounded in $\mathbb{C} \setminus \mathfrak{C}$ $\forall n$, and in the same way the locally uniform boundedness of $\operatorname{Im} h^{(n)}(z)$ in $\mathbb{C} \setminus \mathfrak{C}$ follows. By theorem II.4.1(V), then, the limit function h is again a (ν, μ)-solution in $\mathbb{C} \setminus \mathfrak{C}$. Analogously we deal with $\mathrm{i} \int_{\mathfrak{C}} u(t) \, \mathrm{d}_t H^-(z, t)$. Theorem 3.3 is proved.

We now want to derive saltus conditions for generalized Cauchy integrals 3.3(I), which are analogous with that for usual Cauchy integrals (cf., e.g. [L/S], p. 291). For the sake of simplicity we restrict ourselves to the following situation:

3.6 (a) *Let the curve $\mathfrak{C}$ be simply closed (i.e. without multiple points, apart from the initial and terminal points coinciding) and positively oriented, and let the parameter representation $t(s)$ with the arc length s on $\mathfrak{C}$ as parameter, be continuously differentiable at every point of $\mathfrak{C}$.*

(b) *The function $f(t(s)) = u(t(s)) + \mathrm{i}v(t(s))$, u and v real-valued, is to be continuously differentiable with respect to s in $(0, |\mathfrak{C}|)$ for any parameter representation $t(s)$ just mentioned (i.e. for any choice of the initial point of $\mathfrak{C}$), $|\mathfrak{C}|$ being the total length of $\mathfrak{C}$.*

Then we have:

3.7 *At each point $z_0 \in \mathfrak{C}$ there exists the Cauchy principal value*

$$F(z_0) := \frac{1}{2\pi\mathrm{i}} \lim_{\delta \to} \int_{\mathfrak{C}_\delta} u(t) \, \mathrm{d}_t H^-(z_0, t) + \mathrm{i}v(t) \, \mathrm{d}_t H(z_0, t).$$

If we here take z_0 to be the initial = terminal point of $\mathfrak{C}$ (i.e. $z_0 = t(0) = t(|\mathfrak{C}|)$) then $\mathfrak{C}_\delta$ means that part of $\mathfrak{C}$ which joins $A_\delta = t(s_{1,\delta})$ (where $s_{1,\delta} = \min\{s : |t(s) - z_0| = \delta\}$) and $B_\delta = t(s_{2,\delta})$ ($s_{2,\delta} = \max\{s : |t(s) - z_0| = \delta\}$) along increasing values s.

Proof. We have

$$\lim_{\delta\to 0}\int_{\mathfrak{C}_\delta} v(t)\,\mathrm{d}_t H(z_0, t) = \lim_{\delta\to 0}\Big[v(B_\delta)\,H(z_0, B_\delta) - v(A_\delta)\,H(z_0, A_\delta)$$

$$-\int_{s_{1,\delta}}^{s_{2,\delta}} H(z_0, t(s))\frac{\mathrm{d}v(t(s))}{\mathrm{d}s}\,\mathrm{d}s\Big].$$

Now by 2.1 and IV.3.18 (IV.4.20) the imaginary part of $H(z_0, t(s))$ is continuous at $s = 0$ from the right and at $s = |\mathfrak{C}|$ from the left, and the real part, by 2.1 and IV.3.17, satisfies

3.8 $$|\mathrm{Re}\, H(z_0, t(s))| \leqq \text{constant}\cdot|\log|t(s) - z_0|| + \text{constant},$$

where $|t(s) - z_0| = s + o(s)$ as $s \to +0$ and $|t(s) - z_0| = |\mathfrak{C}| - s + o(|\mathfrak{C}| - s)$ as $s \to |\mathfrak{C}| - 0$. Hence $H(z_0, t(s)) \in L_1((0, |\mathfrak{C}|))$, and, since $\dfrac{\mathrm{d}v(t(s))}{\mathrm{d}s}$ is bounded in $(0, |\mathfrak{C}|)$, by the Lebesgue convergence theorem there exists

3.9 $$\lim_{\delta\to 0}\int_{\mathfrak{C}_\delta} H(z_0, t)\,\mathrm{d}_t v(t) =: {}^{\mathfrak{C}}\!\!\int_{z_0}^{z_0} H(z_0, t)\,\mathrm{d}_t v(t).$$

Furthermore,

$$\mathrm{Re}\lim_{\delta\to 0}\,[v(B_\delta)\,H(z_0, B_\delta) - v(A_\delta)\,H(z_0, A_\delta)]$$

$$= \lim_{\delta\to 0}\mathrm{Re}\,[v(B_\delta)\,M(B_\delta, z_0) - v(A_\delta)\,M(A_\delta, z_0)]$$

$$= \lim_{\delta\to 0} v(z_0)\cdot\mathrm{Re}\,(M(B_\delta, z_0) - M(A_\delta, z_0))$$

$$= \lim_{\delta\to 0} v(z_0)\cdot\mathrm{Re}\,(\alpha - \bar{\alpha}b)(\log|\delta\,\mathrm{e}^{\mathrm{i}\varphi_\delta} + \ell\delta\,\mathrm{e}^{-\mathrm{i}\varphi_\delta}| - \log|-\delta\,\mathrm{e}^{\mathrm{i}\varphi_\delta'} - \ell\delta\,\mathrm{e}^{-\mathrm{i}\varphi_\delta'}|)$$

where $\delta\,\mathrm{e}^{\mathrm{i}\varphi_\delta} = B_\delta - z_0$, $-\delta\,\mathrm{e}^{\mathrm{i}\varphi_\delta'} = A_\delta - z_0$, $\lim_{\delta\to 0}\mathrm{e}^{\mathrm{i}\varphi_\delta} = \lim_{\delta\to 0}\mathrm{e}^{\mathrm{i}\varphi_\delta'} = \lim_{s\to 0} -t'(s) =: \mathrm{e}^{\mathrm{i}\varphi}$, α, ℓ, b as in IV.3.17. Hence

$$\mathrm{Re}\lim_{\delta\to 0}\,[v(B_\delta)\,H(z_0, B_\delta) - v(A_\delta)\,H(z_0, A_\delta)] = 0.$$

In the same manner for the imaginary part of this limit we obtain

$$\mathrm{Im}\lim_{\delta\to 0}\,[v(B_\delta)\,N(B_\delta, z_0) - v(A_\delta)\,N(A_\delta, z_0)]$$

$$= v(z_0)\cdot\lim_{\delta\to 0}\,[\arg(\delta\,\mathrm{e}^{\mathrm{i}\varphi_\delta} + \ell\delta\,\mathrm{e}^{-\mathrm{i}\varphi_\delta}) - \arg(-\delta\,\mathrm{e}^{\mathrm{i}\varphi_\delta'} - \ell\delta\,\mathrm{e}^{-\mathrm{i}\varphi_\delta'})] = v(z_0)\,\pi,$$

thus

3.10 $$\int_{\mathfrak{C}} v(t)\,\mathrm{d}_t H(z_0, t) = \pi v(z_0) - {}^{\mathfrak{C}}\!\!\int_{z_0}^{z_0} H(z_0, t)\,\mathrm{d}_t v(t) \quad \text{for } z_0 \in \mathfrak{C},$$

where we have to understand both integrals in 3.10 to be Cauchy principal values. When replacing μ by $-\mu$ and v by u we obtain from it

3.11 $$\int_{\mathfrak{C}} u(t)\,\mathrm{d}_t H^-(z_0, t) = \pi u(z_0) - {}^{\mathfrak{C}}\!\!\int_{z_0}^{z_0} H^-(z_0, t)\,\mathrm{d}_t u(t) \quad \text{for } z_0 \in \mathfrak{C},$$

where the integral on the right-hand side is defined analogously with 3.9 and exists by the same reasons as the integral on the right-hand side of 3.10. This completes the proof of 3.7. Besides this we have shown

$$\textbf{3.12}\quad F(z_0) = -\frac{1}{2\pi\mathrm{i}} \overset{\mathfrak{C}}{\int_{z_0}^{z_0}} H^-(z_0, t)\, \mathrm{d}_t u(t) + \mathrm{i}H(z_0, t)\, \mathrm{d}_t v(t) + \frac{1}{2} f(z_0)$$

for every $z_0 \in \mathfrak{C}$.

The next theorem embodies the said saltus conditions.

3.13 Theorem. *Let* $\mathfrak{C}$ *and* f *satisfy* 3.6 *and* v, μ *satisfy* 3.1. *Then for each* $z_0 \in \mathfrak{C}$ *and for any sequences* $z_n^+ \to z_0$, $z_n^- \to z_0$, *with* z_n^+ *outside* $\mathfrak{C}$ *and* z_n^- *inside* $\mathfrak{C}$, *there exist the limits*

$$F^+(z_0) := \lim_{n\to\infty} F(z_n^+), \qquad F^-(z_0) := \lim_{n\to\infty} F(z_n^-)$$

satisfying the relations

(I) $F^+(z_0) = F(z_0) + \frac{1}{2}f(z_0)$, $F^-(z_0) = F(z_0) - \frac{1}{2}f(z_0)$.

Proof. For $z \notin \mathfrak{C}$, $z_0 \in \mathfrak{C}$ we have

$$\int_{\mathfrak{C}} u(t)\, \mathrm{d}_t H^-(z, t) + \mathrm{i}v(t)\, \mathrm{d}_t H(z, t)$$

$$= [u(t)\, H^-(z, t) + \mathrm{i}v(t)\, H(z, t)]_{z_0}^{z_0} - \overset{\mathfrak{C}}{\int_{z_0}^{z_0}} H^-(z, t)\, \mathrm{d}_t u(t) + \mathrm{i}H(z, t)\, \mathrm{d}_t v(t)$$

$$= \begin{cases} 2\pi\mathrm{i}f(z_0) & \text{if } z \text{ inside } \mathfrak{C} \\ 0 & \text{if } z \text{ outside } \mathfrak{C} \end{cases}$$

$$- \overset{\mathfrak{C}}{\int_{z_0}^{z_0}} (H^-(z,t) - H^-(z_0,t))\, \mathrm{d}_t u + \mathrm{i}(H(z,t) - H(z_0,t))\, \mathrm{d}_t v + \mathrm{i}\pi(2F(z_0) - f(z_0)),$$

by 3.12. Furthermore,

$$\textbf{3.14}\quad \lim_{z\to z_0} \overset{\mathfrak{C}}{\int_{z_0}^{z_0}} (H^-(z, t) - H(z_0, t))\, \mathrm{d}_t u(t) = \lim_{z\to z_0} \overset{\mathfrak{C}}{\int_{z_0}^{z_0}} (H(z, t) - H(z_0, t))\, \mathrm{d}_t v(t) = 0.$$

Indeed, in part by continuity and in part by definition, for instance, the first limit in 3.14 is

$$\textbf{3.15}\quad = \lim_{z\to z_0} \Big(\lim_{\delta\to 0} \int_{\mathfrak{C}_\delta} (H^-(z, t) - H^-(z_0, t))\, \mathrm{d}_t u(t)\Big),$$

$\mathfrak{C}_\delta$ as in 3.7. On account of IV.4.22 and the conditions on $u(t)$, for every $\varepsilon > 0$ there exists a $\delta_0 > 0$ so that

$$\textbf{3.16}\quad \Big|\int_{\mathfrak{C}_\delta \setminus \mathfrak{C}_{\delta_0}} (H^-(z, t) - H^-(z_0, t))\, \mathrm{d}_t u(t)\Big| < \frac{\varepsilon}{2} \quad \forall \delta \in (0, \delta_0) \text{ and, say,}$$

$$\forall z \in \{|z - z_0| \leqq 1\}.$$

Of course, we also have (for example by the Lebesgue convergence theorem)

$$\textbf{3.17}\quad \lim_{z\to z_0} \int_{\mathfrak{C}_{\delta_0}} (H^-(z, t) - H^-(z_0, t))\, \mathrm{d}_t u(t) = 0,$$

because the integrand is uniformly bounded on $\mathfrak{C}_{\delta_0}$ for every z from $\{|z - z_0| < \delta_0/2\}$ and converges pointwise to zero there. The same considerations also hold for the other limit in 3.14. This gives 3.14 and thus also 3.13.

3.18 ***Remark.*** On account of the relatively strong conditions on f and $\mathfrak{C}$ we do not need any restrictions for approaching $z_0 \in \mathfrak{C}$ by the z_n^+, z_n^-. In particular 3.13(I) holds true if the approach takes place tangentially.

V.4 GENERALIZATION OF THE SCHWARZ AND POISSON FORMULAE

In a domain $G \supseteqq \{|z| \leqq 1\}$ let the coefficients ν_0, μ_0 possess the following properties:

4.1 $\nu_0, \mu_0 \in HL_p(\{|z| \leqq 1\})$, $p > 2$, $|\nu_0(z)| + |\mu_0(z)| \leqq k < 1$ $\forall z \in G$, μ_0 is to be real when $|z| = 1$, and ν_0 is to be equal to zero at $z = 0$ and on $\{|z| = 1\}$.

Now we define new coefficients ν, μ in $\overline{\mathbb{C}}$ by

4.2 $$\nu(z) = \begin{cases} \nu_0(z) & \text{if } |z| < 1, \\ (z/\bar{z})^2\, \overline{\nu_0(1/\bar{z})} & \text{if } |z| \geqq 1, \end{cases} \qquad \mu(z) = \begin{cases} \mu_0(z) & \text{if } |z| < 1, \\ \overline{\mu_0(1/\bar{z})} & \text{if } |z| \geqq 1. \end{cases}$$

Then it holds

4.3 (a) $\nu, \mu \in HL_p\left(\{|z| \leqq 1 - d\} \cup \{|z| = 1\} \cup \left\{\frac{1}{d} \geqq |z| \geqq 1 + d\right\} \cup \{\infty\}\right)$ *with* $p > 2$ *as in* 4.1 *and every* $d \in (0, 1/2)$;

(b) ν, μ *satisfy the symmetry condition*

$$\nu(z) = (z/\bar{z})^2\, \overline{\nu(1/\bar{z})}, \qquad \mu(z) = \overline{\mu(1/\bar{z})}.$$

With these ν, μ let $M(t, z)$, $N(t, z)$, $H(t, z)$, $H^-(t, z)$ be the functions mentioned in 3.2. Further let $w(z)$ be a (ν, μ)-solution in G. An easy calculation shows, then, that

4.4 $\tilde{w}(z) := \overline{w(1/\bar{z})}$ is a (ν, μ)-solution in $\{z\colon 1/\bar{z} \in G\}$.

Hence, by the Cauchy integral formula,

$$w(z) = \frac{1}{2\pi i} \oint_{|t|=1} \operatorname{Re}(w(t)\, d_t N(t, z)) + i \operatorname{Im}(w(t)\, d_t M(t, z))$$

as well as

$$0 = \frac{1}{2\pi i} \oint_{|t|=R} \operatorname{Re}(\tilde{w}(t)\, d_t N(t, z)) + i \operatorname{Im}(\tilde{w}(t)\, d_t M(t, z))$$
$$- \frac{1}{2\pi i} \oint_{|t|=1} \operatorname{Re}(\overline{w(t)}\, d_t N(t, z)) + i \operatorname{Im}(\overline{w(t)}\, d_t M(t, z)),$$

where $R > 1 > |z|$. Substraction of the second equation from the first gives

4.5

$$w(z) = \frac{1}{2\pi i} \oint_{|t|=1} \operatorname{Re}[(w + \bar{w})\, d_t N(t, z)] + i \operatorname{Im}[(w + \bar{w})\, d_t M(t, z)] - K(R, z),$$

where

$$K(R, z) := \frac{1}{2\pi i} \oint_{|t|=R} \operatorname{Re}(\tilde{w}(t)\, d_t N(t, z)) + i \operatorname{Im}(\tilde{w}(t)\, d_t M(t, z))$$

$$= \frac{1}{2\pi i} [\operatorname{Re}(2\pi i \tilde{w}(R)) + i \operatorname{Im}(2\pi i \tilde{w}(R))]$$

$$- \frac{1}{2\pi i} \oint_{R,\, |t|=R}^{R} \operatorname{Re}[N(t, z)\, d_t \tilde{w}(t)] + i \operatorname{Im}[M(t, z)\, d_t \tilde{w}(t)].$$

Obviously $\lim_{R\to\infty} \tilde{w}(R) = \overline{w(0)}$ and

$$\lim_{R\to\infty} \oint_{R,\, |t|=R}^{R} N(t, z)\, d_t \tilde{w}(t) = \lim_{r\to 0} \oint_{r,\, |s|=r}^{r} N(1/\bar{s}, z)\, d_s \overline{w(s)} = 0$$

because of the boundedness of the derivatives of w in $\{|s| \leqq 1\}$ and IV.4.17. By the same reason also

$$\lim_{R\to\infty} \oint_{R,\, |t|=R}^{R} M(t, z)\, d_t \tilde{w}(t) = 0,$$

and when putting $u = \operatorname{Re} w$, $v = \operatorname{Im} w$, from this we obtain

4.6 $$w(z) = \frac{1}{2\pi i} \oint_{|t|=1} u(t) \cdot 2\operatorname{Re} d_t N(t, z) + iu(t) \cdot 2\operatorname{Im} d_t M(t, z) - \overline{w(0)}.$$

For $z = 0$ from 4.6 we first infer

4.7 $$u(0) = \frac{1}{2\pi i} \oint_{|t|=1} u(t) [\operatorname{Re} d_t N(t, 0) + i \operatorname{Im} d_t M(t, 0)].$$

Inserting this into 4.6 gives the generalized Schwarz formula

4.8 $$w(z) = iv(0) + \frac{1}{2\pi i} \oint_{|t|=1} u(t) [\operatorname{Re} d_t (2N(t, z) - N(t, 0)) + i \operatorname{Im} d_t (2M(t, z) - M(t, 0))].$$

By separating the real and imaginary parts in 4.8 we finally obtain the generalized Poisson integral formula

4.9 $$u(z) = \frac{1}{2\pi} \oint_{|t|=1} u(t)\, d_t \operatorname{Im}(2M(t, z) - M(t, 0)), \quad |z| < 1.$$

Now let $u(t)$ be any piecewise continuous, real-valued function. By theorem 3.3 the integral on the right-hand side of 4.9 represents the real part of a (ν, μ)-solution in $\mathbb{C} \setminus \{|z| = 1\}$. We want to show that this generalized Poisson integral, analogously with the classical case, yields the solution of the first boundary value problem for the disk, namely to give the real part of a (ν, μ)-solution in a disk (ν, μ subject to 4.1), taking prescribed boundary values. Simultaneously this embodies the solution of this boundary value problem for a second order elliptic differential equation of the type

4.10
$$(au_x + bu_y)_x + (cu_x + du_y)_y = 0,$$

where a, b, c, d correspond to ν, μ according to II.1.13.

To this end we first prove

4.11 $g(t) := M(t, z) - M(t, 0) - \overline{M(1/\bar{t}, 1/\bar{z})} =$ *constant as a function of* t *with arbitrary but fixed* $z \neq 0$, $z \neq \infty$.

Proof. The function $g(t)$ can be defined uniquely in $\mathbb{C} \setminus \{0, z\}$, and on account of 4.3 it is a $(\nu, \bar{\mu})$-solution there (together with $\nu, \mu - \nu, \bar{\mu}$ also satisfy the symmetry condition 4.3(b)). At the points $t = 0$, $t = z$ and for $t \to \infty$ $g(t)$ has a certain growth less than $|t|^{-\gamma}$, $|t - z|^{-\gamma}$ and $|t|^{\gamma}$, respectively, for every positive γ, because of the asymptotic expansions IV.4.20, IV.4.14 (cf. also IV.4.16, IV.4.18). Then 0, z can be neither essential singularities (by the representation theorem and the Hölder continuity of quasiconformal mappings) nor poles (by theorem III.3.2), hence they are removable. Thus, by theorem III.2.7, $g(t) =$ constant, which we had to prove.

If $z \neq 0$, $|z| < 1$, in place of 4.9 we also could write

4.12
$$u(z) = \frac{1}{2\pi} \oint_{|t|=1} u(t)\, \mathrm{d}_t U(t, z) \quad \text{where} \quad U(t, z) := \operatorname{Im}\left(M(t, z) + \overline{M(t, 1/\bar{z})}\right).$$

Next we consider piecewise constant boundary values in the simplest case.

4.13 Lemma. *Let* γ *be a subarc of the unit circle,* $\gamma = \{e^{i\varphi}: \varphi_1 < \varphi < \varphi_2,$ $0 < \varphi_2 - \varphi_1 < 2\pi$, $\varphi_2 \in (0, 2\pi]\}$, *and* $h(t)$ *be* $= 1$ *if* $t \in \gamma$ *and* $= 0$ *otherwise. Then*

$$\lim_{\substack{z \to z_0 \\ |z| < 1}} \frac{1}{2\pi} \oint_{|t|=1} h(t)\, \mathrm{d}_t U(t, z) = h(z_0) \quad \textit{for every } z_0 \in \{|z| = 1\} \setminus \{e^{i\varphi_1}, e^{i\varphi_2}\}.$$

If z *tends to* $e^{i\varphi_1}$ *or* $e^{i\varphi_2}$ *inside the unit disk, then all accumulation points of the Poisson integral are situated in the interval* $[0, 1]$, *and every* $a \in [0, 1]$ *occurs as an accumulation point of a suitable sequence of* $z \to e^{i\varphi_j}$, $|z| < 1$, $j = 1, 2$.

Proof. In order to specify the occuring argument functions we think of $\mathbb{C}$ to be slit along γ and from $e^{i\varphi_1}$ to infinity along a ray (say, tangential to the unit circle at $e^{i\varphi_1}$ and forming the angle π with γ). Let us consider the case $z \to e^{i\varphi_2} =: t_2$, $|z| < 1$. We have (first with z fixed)

$$\oint_{|t|=1} h(t)\, \mathrm{d}_t U(t, z) = \int_{\gamma} \mathrm{d}_t \operatorname{Im}\left(M(t, z) - M\left(t, \frac{1}{\bar{z}}\right)\right)$$

$$= \int_{\gamma} \mathrm{d}_t \operatorname{Im}\left(H^-(z, t) - H^-\left(\frac{1}{\bar{z}}, t\right)\right)$$

$$= \operatorname{Im}\left[H^-(z, t_2) - H^-\left(\frac{1}{\bar{z}}, t_2\right)\right] - \operatorname{Im}\left[H^-(z, t_1) - H^-\left(\frac{1}{\bar{z}}, t_1\right)\right],$$

where $t_1 := e^{i\varphi_1}$.

According to our cutting $\mathbb{C}$ open it comes

$$\lim_{z \to t_2} \left(H^-(z, t_1) - H^-\left(\frac{1}{\bar{z}}, t_1\right) \right) = 0 \quad \text{and}$$

$$0 \leqq \lim_{z \to t_2} \left[\arg(z - t_2) - \arg\left(\frac{1}{\bar{z}} - t_2\right) \right] \leqq 2\pi.$$

Hence, by IV.4.20 (with μ replaced by $-\mu$ and observing $\ell = \ell(t_2) = 0$, because $\nu = 0$ on $\{|t| = 1\}$), it holds

$$0 \leqq \liminf_{\substack{z \to t_2 \\ |z| < 1}} \operatorname{Im} \left[H^-(z, t_2) - H^-\left(\frac{1}{\bar{z}}, t_2\right) \right]$$

$$\leqq \limsup_{\substack{z \to t_2 \\ |z| < 1}} \operatorname{Im} \left[H^-(z, t_2) - H^-\left(\frac{1}{\bar{z}}, t_2\right) \right] \leqq 2\pi,$$

and if $\lim\limits_{\substack{z \to t_2 \\ |z| < 1}} \arg(z - t_2) =: \psi \in \left[\varphi_2 + \frac{\pi}{2}, \varphi_2 + \frac{3\pi}{2} \right]$, then

$$\lim_{\substack{z \to t_2 \\ |z| < 1}} \operatorname{Im} \left(H^-(z, t_2) - H^-\left(\frac{1}{\bar{z}}, t_2\right) \right) = 2\psi - 2\varphi_2 - \pi \in [0, 2\pi];$$

note that $1/\bar{z}$ preserves (the absolute value of) angles. In the same way the assertion follows when $z \to e^{i\varphi_1} = t_1$.

Next let $z_0 \in \{|z| = 1\} \setminus \bar{\gamma}$. As we have just shown (but now simpler) it arises

4.14
$$\lim_{z \to z_0} \oint_{|t|=1} h(t) \, d_t U(t, z)$$

$$= \lim_{z \to z_0} \operatorname{Im} \left[H^-(z, t_2) - H^-\left(\frac{1}{\bar{z}}, t_2\right) - \left(H^-(z, t_1) - H^-\left(\frac{1}{\bar{z}}, t_1\right) \right) \right] = 0.$$

Finally, let $z_0 \in \gamma$. We consider the function $h^*(t) = 1 - h(t)$. As proved just now, we have

4.15
$$\lim_{z \to z_0} \oint_{|t|=1} h^*(t) \, d_t U(t, z) = \lim_{z \to z_0} \oint_{|t|=1} (1 - h(t)) \, d_t U(t, z) = 0$$

(obviously it is no matter that at present $h^* = 1$ on a closed (in place of an open) arc and $h^* = 0$ elsewhere on the unit circle). Besides, for the (ν, μ)-solution $w(z) \equiv 1$ we clearly have

4.16
$$\frac{1}{2\pi} \oint_{|t|=1} 1 \cdot d_t U(t, z) = 1 \quad \forall z \text{ from the unit disk,}$$

according to 4.8, 4.9. Together with 4.15 this gives

4.17
$$\lim_{\substack{z \to z_0 \\ |z| < 1}} \frac{1}{2\pi} \oint_{|t|=1} h(t) \, d_t U(t, z) = 1 \quad \text{if } z_0 \in \gamma.$$

The proof of lemma 4.13 is complete.

4.18 *The (generalized) Poisson kernel* $\dfrac{\mathrm{d}U(\mathrm{e}^{\mathrm{i}\varphi}, z)}{\mathrm{d}\varphi}$ *is non-negative for every* $\varphi \in [0, 2\pi]$ *and for each* z *with* $|z| < 1$ (*for* $z = 0$ *put* $U(t, 0) = \operatorname{Im} M(t, 0)$).

Proof. Suppose, there is a z_1, $|z_1| < 1$, and a φ_0 so that

$$\left.\frac{\mathrm{d}U(\mathrm{e}^{\mathrm{i}\varphi}, z_1)}{\mathrm{d}\varphi}\right|_{\varphi=\varphi_0} < 0. \tag{4.19}$$

Since $\nu, \mu \in HL_p(\{|t| = 1\})$, $\dfrac{\mathrm{d}U(\mathrm{e}^{\mathrm{i}\varphi}, z_1)}{\mathrm{d}\varphi}$ is continuous on $[0, 2\pi]$, by theorem II.5.2(II). Hence there is an arc γ such that

$$\frac{\mathrm{d}U(\mathrm{e}^{\mathrm{i}\varphi}, z_1)}{\mathrm{d}\varphi} < 0 \quad \text{when } \mathrm{e}^{\mathrm{i}\varphi} \in \gamma.$$

Let again $h(t) = 1$ if $t \in \gamma$ and $h(t) = 0$ else. Then

$$\tilde{h}(z) := \frac{1}{2\pi} \oint_{|t|=1} h(t)\, \mathrm{d}_t U(t, z) = \frac{1}{2\pi} \int_0^{2\pi} h(t) \frac{\mathrm{d}U(\mathrm{e}^{\mathrm{i}\varphi}, z)}{\mathrm{d}\varphi}\, \mathrm{d}\varphi$$

is less than 0 at $z = z_1$. On the other hand, by 4.13

$$\liminf_{|z| \to 1-0} \tilde{h}(z) = 0. \tag{4.20}$$

However, by theorem 3.3 (cf. also 4.9, 4.11, 4.12, 2.5), $h(z)$ is the real part of a (ν, μ)-solution in $\{|z| < 1\}$. Thus, 4.20 and $h(z_1) < 0$ is in contradiction with the maximum principle III.2.4 (applied to $-h(z)$). Assertion 4.18 is proved.

Now we are able to show that the Poisson integral indeed solves the boundary value problem mentioned above. We have the following

4.21 Theorem. *Let $u(t)$ be defined, real-valued and continuous, up to a finite number of jumps, on the unit circle.*

(I) *For the Poisson integral of $u(t)$*

$$\frac{1}{2\pi} \oint_{|t|=1} u(t)\, \mathrm{d}_t U(t, z) =: U(z)$$

in every continuity point z_0 of $u(t)$ it holds

$$\lim_{\substack{z \to z_0 \\ |z| < 1}} U(z) = u(z_0).$$

(II) *Let $u^+(z_0)$, $u^-(z_0)$ be the one-sided limits of $u(t)$ as $t \to z_0$, where z_0 is a discontinuity point of $u(t)$, $u^-(z_0) < u^+(z_0)$. Then*

$$u^-(z_0) = \liminf_{\substack{z \to z_0 \\ |z| < 1}} U(z), \qquad u^+(z_0) = \limsup_{\substack{z \to z_0 \\ |z| < 1}} U(z),$$

and for every $a \in [u^-(z_0), u^+(z_0)]$ *there exists a sequence* $z_n \to z_0$ *as* $n \to \infty$, $|z_n| < 1\ \forall n$, *such that*

$$\lim_{n \to \infty} U(z_n) = a.$$

Proof. By 4.16, 4.18 $|U(z) - u(z_0)| \leqq \frac{1}{2\pi} \oint\limits_{|t|=1} |u(t) - u(z_0)|\, \mathrm{d}_t U(t, z)$. If $z_0 = \mathrm{e}^{\mathrm{i}\varphi_0}$ is a continuity point of $u(t)$, there exists an arc $\gamma = \{\mathrm{e}^{\mathrm{i}\varphi}: |\varphi - \varphi_0| < \delta\}$ so that $|u(t) - u(z_0)| < \varepsilon/2\ \forall t \in \gamma$, hence

$$0 \leqq \frac{1}{2\pi} \int\limits_{\gamma} |u(t) - u(z_0)|\, \mathrm{d}_t U(t, z) \leqq \frac{\varepsilon}{4\pi} \oint\limits_{|t|=1} \mathrm{d}_t U(t, z) = \varepsilon/2$$

for every z within the unit disk. Let M be a bound of $|u(t)|$ on $\{|t| = 1\}$. Then

$$\frac{1}{2\pi} \int\limits_{\{|t|=1\}\setminus\gamma} |u(t) - u(z_0)|\, \mathrm{d}_t U(t, z) \leqq \frac{2M}{2\pi} \int\limits_{\{|t|=1\}\setminus\gamma} \mathrm{d}_t U(t, z) \to 0$$

as $z \to z_0$, by lemma 4.13. Thus 4.21 (I) is proved.

Now let $z_0 = \mathrm{e}^{\mathrm{i}\varphi_0}$ be a discontinuity point of $u(t)$. By adding a suitable jump function $h(t) = \text{constant} \neq 0$ on a corresponding arc γ and $h(t) = 0$ on $\{|t| = 1\} \setminus \gamma$, there arises a function $u(t) + h(t)$ continuous at z_0. From 4.21 (I), applied to the Poisson integral of $u(t) + h(t)$, and from 4.13, applied to the Poisson integral of $h(t)$, we obtain 4.21 (II), which completes the proof of theorem 4.21.

Chapter VI

VARIATIONAL METHODS FOR SCHLICHT (ν, μ)-SOLUTIONS

The theory of extremal problems for conformal mappings can, in a natural way, be generalized in different directions. One direction—without doubt the main direction of all possible generalizations—arose by simply considering certain extremal problems, which had been treated hitherto only for conformal mappings, also for quasiconformal mappings being subject (apart from certain normalizations) solely to a dilatation condition $D_f(z) \leqq K(z)$ (cf. I.1.1 (III′)), where $K(z)$ is a prescribed measurable bounded function $\geqq 1$. (Concerning literature, let it be enough to refer to the bibliography of [K1] and [S4]. Concerning an extension of this problem cf. [S/S1, 2].) A second generalization (suggested in [K2], § 6, and treated in more detail for the first time in [R2, 3], cf. also a related problem in [G/S]) consists of replacing the Cauchy–Riemann system by any fixed (ν, μ)-system and seeking to characterize the extremal functions in suitably normalized classes of (ν, μ)-solutions.

To start with, nothing is known about the nature of such extremal mappings. At this stage the only possibility to attack the problem appears to be a suitable variational method. Accordingly (as the first task in this complex of problems), we have to look for (below called $\varkappa$-conformal) boundary variations being analogues to the Schiffer conformal boundary variations and not leading beyond the mapping class in question. Indeed we shall show that such boundary variations do exist, namely as transformations of the Schiffer conformal boundary variations.

Here we restrict ourselves to classes of schlicht (ν, μ)-solutions, f being defined near infinity and there normalized by an asymptotic expansion with $f(\infty) = \infty$, as we have considered it in III.4; these f, however, need not satisfy the (ν, μ)-system in a boundary area, but are there subject only to a dilatation condition of the above mentioned kind. (This simultaneously means a certain synthesis of both the above mentioned directions of generalizing extremal problems in conformal mapping; in particular that boundary area could be the whole of G apart from a neighbourhood of infinity needed for normalization purposes.) The presence of this boundary area not only permits but also requires certain inner variations for characterizing the extremal mappings.

In a sense the type of variations determines the type of extremal problems. In this way our variations lead to extremal problems which are again (analogously with the conformal case) ruled by (generalized) quadratic differentials, being transformations of corresponding quadratic differentials in the conformal case (cf. the examples in section VI.7 as well as a 'similarity principle' in [R2], which could be interpreted to mean a transmission of the Teichmüller principle relating the features of certain extremal problems with the kind and position of singularities of the corresponding quadratic differentials). These extremal problems, respectively the corresponding functionals, shall be called extremal problems, respectively functionals, of Grötzsch-Teichmüller type. Via Schiffer's lemma and variational considerations analogous with [S3], [R1], their definition implies the validity of a 'main theorem' (6.4 below, cf. [R1]).

VI.1 PRELIMINARIES

Henceforth let G be a domain of the extended plane with $\infty \in G$, $D_0 := \{|z| \geqq R_0\} \subset G$, R_0 a positive number. The functions $\nu(z)$, $\mu(z)$ are to be defined and measurable in $\overline{\mathbb{C}}$ and to satisfy

1.1 $|\nu(z)| + |\mu(z)| \leqq \text{constant} =: k < 1 \quad \forall z \in \overline{\mathbb{C}}, \qquad \nu(z) = \mu(z) = 0 \quad \forall z \notin G,$

$|\nu(z) - \nu(\infty)| + |\mu(z) - \mu(\infty)| \in L_q, \quad 1 < q < 2.$

Further, let D be a subset of G such that,

1.2 $$D_0 \Subset D \Subset G.$$

1.3 ***Definition.*** (I) Let $\mathfrak{Q}(\nu, \mu)$ be the set of all quasiconformal mappings f of G satisfying

(1) $|f_{\bar{z}}| \leqq (|\nu(z)| + |\mu(z)|)\,|f_z|$ a.e. in G,

(2) $f_{\bar{z}} = \nu(z) f_z + \mu(z) \overline{f_z}$ a.e. in D,

(3) $f(z) = z + (\nu(\infty) + \mu(\infty))\,\bar{z} + \varepsilon_f(z)$ with $\lim_{z \to \infty} \varepsilon_f(z) = 0$.

(II) A schlicht (ν, μ)-solution in a neighbourhood of infinity, admitting an expansion (3), is called hydrodynamically normalized (with respect to this (ν, μ)-system).

By the theorems III.4.28, III.4.20 $\mathfrak{C}(\nu, \mu)$ is non-empty and compact (in itself) in the sense of III.4.19 (the agreement of III.4.19 is to hold also furthermore).

On account of the compactness of $\mathfrak{Q}(\nu, \mu)$ there exists a positive constant R^* so that

1.4 $$\{|w| \geqq R^*\} \subseteqq f(D_0) \quad \forall f \in \mathfrak{Q}(\nu, \mu).$$

Now for any fixed $f \in \mathfrak{Q}(\nu, \mu)$ let

1.5 $$\varkappa(w) = \varkappa_f(w) := \frac{-\mu(z(w))}{1 + |\mu(z(w))|^2 - |\nu(z(w))|^2} \quad \text{if } w \in f(D),$$
$$\varkappa(w) = 0 \ \text{ otherwise,}$$

where $z(w)$ means the inverse mapping of $f(z)$. For every such $\varkappa$ we have

1.6 $$|\varkappa(w)| \leqq \frac{1}{2}\left(1 - \frac{(1-k)^2}{1+k^2}\right)$$

(note $|\nu|^2 \leqq (1 - k + |\nu|)^2 - (1 - k)^2 \leqq (1 - |\mu|)^2 - (1 - k)^2$).

For every schlicht $(\varkappa, -\varkappa)$-solution $\mathfrak{F}$ in $G_w = f(G)$, admitting an expansion

1.7 $$\mathfrak{F}(w) = w + \varepsilon_{\mathfrak{F}}(w) \quad \text{with} \lim_{w \to \infty} \varepsilon_{\mathfrak{F}}(w) = 0$$

(that is, $\mathfrak{F}$ is hydrodynamically normalized with respect to the $(\varkappa, -\varkappa)$-system), it holds

1.8 $$f^*(z) := \mathfrak{F} \circ f(z) \in \mathfrak{Q}(\nu, \mu),$$

as we can show by an easy computation (immediately or by using the considerations from II.3). (Of course, it makes no sense for $\mathfrak{F}$ to take the schlicht $(\varkappa, -\varkappa)$-solution in $\mathbb{C}$ with normalization 1.7; this $\mathfrak{F}$ is the identical mapping.)

We now want to construct a hydrodynamically normalized schlicht $(\varkappa, -\varkappa)$-solution $\mathfrak{F}$ (additionally depending on a parameter λ, $\mathfrak{F} = \mathfrak{F}(\cdot, \lambda)$) by means of which $f \in \mathfrak{Q}(\nu, \mu)$ can be varied. For this we suppose:

1.9 *G has at least one non-degenerate boundary component (i.e. a boundary component containing at least two distinct points).*

Then also $\partial f(G)$ has at least one non-degenerate boundary component (otherwise we would obtain a contradiction, by means of the theorem of Riemann in conjunction with the representation theorem, applied to the inverse mapping of $f \in \mathfrak{C}(\nu, \mu)$, cf. also I.2.33).

Let $\mathfrak{c}$ be a continuum $\subset \partial f(G)$ (f arbitrary but fixed $\in \mathfrak{Q}(\nu, \mu)$), $\mathfrak{c}$ containing at least two distinct points, let the complement of $\mathfrak{c}$ be a domain $G_{\mathfrak{c}}$, and let w_0 be a point of $\mathfrak{c}$. By $\boldsymbol{F}(\mathfrak{c})$ we denote the set of all conformal hydrodynamically normalized mappings of $G_{\mathfrak{c}}$. In $\boldsymbol{F}(\mathfrak{c})$ there are sequences of mappings F_n so that (cf. [S2])

1.10 (I) $$\lambda_n := \frac{1}{2\pi i} \oint_{|w| = R^*} F_n(w)\, dw \to 0 \quad \text{as } n \to \infty, \ \lambda_n \neq 0 \quad \forall n = 1, 2, \ldots,$$

(II) *for* $o(\lambda_n) := F_n(w) - w - \dfrac{\lambda_n}{w - w_0}$ *we have* $o(\lambda_n)/\lambda_n \to 0$ *uniformly in each* $G' \Subset G_{\mathfrak{c}}$ *as* $n \to \infty$.

If any $F \in \boldsymbol{F}(\mathfrak{c})$ acts as an element of such a sequence, we will call it a conformal boundary variation of $G_w = f(G)$ with respect to $\mathfrak{c}$ and w_0, and we will denote it by $F(w, \lambda_n)$. In future we usually omit the index n of λ_n. Nevertheless '$\lambda \to 0$' only means that we are given a sequence of conformal boundary variations $F(w, \lambda_n)$ and λ runs through the null sequence of the λ_n. Accordingly, for instance, the terminology 'sufficiently small λ' only means that n is sufficiently big in the null sequence λ_n at the time.

By means of a correction mapping $\mathfrak{v}$ to be constructed next, the $F(w, \lambda)$ are to be transformed into $(\varkappa, -\varkappa)$-solutions $\mathfrak{F}(w, \lambda)$.

Let $w(\zeta)$ be the inverse mapping of the conformal boundary variation $F(w, \lambda) \in \boldsymbol{F}(\mathfrak{c})$. With $\varkappa$ defined in 1.5 let

1.11 $$\varkappa_\lambda(\zeta) = \varkappa(w(\zeta)) \quad \text{if } w(\zeta) \in G_w, \ \varkappa_\lambda(\zeta) = 0 \ \text{otherwise},$$

$$\varkappa_\lambda^*(\zeta) = \varkappa_\lambda(\zeta)\left(\frac{\overline{w'(\zeta)}}{w'(\zeta)} - 1\right) \quad \text{if } w(\zeta) \in G_w, \ \varkappa_\lambda^*(\zeta) = 0 \ \text{otherwise}.$$

1.12 *There exists just one hydrodynamically normalized schlicht solution of the system*

$$\text{(I)} \ \mathfrak{v}_{\bar{\zeta}} = \varkappa_\lambda(\zeta)\frac{\overline{w'(\zeta)}}{w'(\zeta)}\mathfrak{v}_\zeta - \varkappa_\lambda(\zeta)\,\overline{\mathfrak{v}_\zeta} = 2i\varkappa_\lambda(\zeta)\,\mathrm{Im}\,\mathfrak{v}_\zeta + \varkappa_\lambda^*(\zeta)\,\mathfrak{v}_\zeta \ \textit{in } \mathbb{C}.$$

Proof. The uniqueness of $\mathfrak{v}(\zeta)$ at once follows since it is hydrodynamically normalized and also from Liouville's theorem III.2.6. In the case of the system 1.12(I) with $\varkappa_\lambda(\infty) = \varkappa(\infty)$, $\varkappa_\lambda^*(\infty) = 0$ the hydrodynamical normalization here reads

1.13 $$\mathfrak{v}(\zeta) = \zeta + \varepsilon_{\mathfrak{v}}(\zeta) \quad \text{with} \lim_{\zeta \to \infty} \varepsilon_{\mathfrak{v}}(\zeta) = 0.$$

In order to prove existence (which does not immediately follow from III.4.28) we consider the equation

1.14 $$h = T\varkappa_\lambda^* + T(2i\varkappa_\lambda\,\mathrm{Im}\,h + \varkappa_\lambda^* h).$$

Since $\varkappa_\lambda^*(\zeta) \in L_s$ for every $s > 1$ and

1.15 $$\|T(2i\varkappa_\lambda\,\mathrm{Im}\,h + \varkappa_\lambda^* h)\|_{L_p} \leqq 2C(p)\,\|\varkappa\|_{L_\infty}\,\|h\|_{L_p}$$
$$\leqq C(p)\left(1 - \frac{(1-k)^2}{1+k^2}\right)\|h\|_{L_p},$$

there exists an $\varepsilon_0 \in (0, 1)$ so that

1.16
$$C(p)\left(1 - \frac{(1-k)^2}{1+k^2}\right) \leqq 1 - \frac{(1-k)^2}{2(1+k^2)} =: k' < 1 \quad \forall p \in [2 - \varepsilon_0, 2 + \varepsilon_0].$$

The equation 1.14 is real-linear. Hence, by the Banach fixed point theorem there exists a unique solution in $L^{p,q}$ with $p \in (2, 2 + \varepsilon_0]$, $q \in [2 - \varepsilon_0, 2)$. The desired $\mathfrak{v}(\zeta)$ then is

1.17 $$\mathfrak{v}(\zeta) = \zeta + P(2i\varkappa_\lambda \operatorname{Im} h + \varkappa_\lambda^*(h + 1))(\zeta).$$

Indeed, this $\mathfrak{v}(\zeta)$ is a solution of 1.12(I), by theorem 0.4.19 and 1.14 above. From the representation theorem, the argument principle and theorem 0.4.3 (IV) the schlichtness and hydrodynamical normalization follow. Assertion 1.12 is proved.

1.18 *When $\lambda \to 0$, $\mathfrak{v}(\zeta) \to \zeta$ uniformly in $\mathbb{C}$ as well as $\mathfrak{v}_\zeta - 1 \to 0$ in L_p for every $p \in [2 - \varepsilon_0, 2 + \varepsilon_0]$.*

Proof. The assertion $h = \mathfrak{v}_\zeta - 1 \to 0$ in L_p follows from 1.14, 1.15 and from the continuity of the T-operator in L_p, once we know that $\|\varkappa_\lambda^*\|_{L_p}$ tends to zero for every $p \in [2 - \varepsilon_0, 2 + \varepsilon_0]$. As to the latter,

$$\|\varkappa_\lambda^*\|_{L_p}^p = \iint\limits_{f(D)} |\varkappa(w)|^p \left|\frac{F'(w, \lambda)}{\overline{F'(w, \lambda)}} - 1\right|^p |F'(w, \lambda)|^2 \, d\sigma_w$$
$$\leqq \iint\limits_{f(D)} |F'(w, \lambda) - \overline{F'(w, \lambda)}| \cdot |F'(w, \lambda)|^{2-p} \, d\sigma_w.$$

Observing $f(D) \Subset G_w \subset G_{\mathfrak{c}}$ and 1.10(II) we see that the last expression tends to zero as $\lambda \to 0$, if

1.19 $$\iint\limits_{G'} |F'(w, \lambda) - 1|^p \, d\sigma_w \to 0 \text{ for every } p > 1 \text{ and every } G' \Subset G_w \text{ as } \lambda \to 0.$$

This can be shown as follows. By 1.4, the $F(w, \lambda)$ in particular are conformal hydrodynamically normalized mappings of $\{|w| \geqq R^*\}$, i.e.

1.20 $$F(w, \lambda) = w + \frac{\lambda}{w} + \frac{a_2}{w^2} + \frac{a_3}{w^3} + \dots \quad \text{for } |w| \geqq R^*.$$

Here the coefficients a_j are given by

$$a_j = \frac{1}{2\pi i} \oint\limits_{|w| = R^*} F(w, \lambda) \, w^{j-1} \, dw = \frac{1}{2\pi i} \oint\limits_{|w| = R^*} \left(w + \frac{\lambda}{w - w_0} + o(\lambda)\right) w^{j-1} \, dw$$
$$= \lambda w_0^{j-1} + o(\lambda) \, R^{*j}$$

where the last $o(\lambda)$ obviously is dominated by an $o(\lambda)$ which is independent of j. Hence, by 1.10(II), for every sequence of boundary variations $F(w, \lambda)$ with respect to $\mathfrak{c}$ and w_0 there exists a constant M so that

1.21 $$|a_j| \leqq |\lambda| \cdot M \cdot R^{*j} \quad \forall j = 2, 3, \dots$$

By 1.20, 1.21, 1.10(II), in every $G' \Subset G_w$ there exists an integrable majorant of $|F'(w, \lambda) - 1|$. Besides, of course, $F'(w, \lambda) \to 1$ pointwise in $G_{\mathfrak{c}}$ as $\lambda \to 0$. By the Lebesgue convergence theorem, 1.19 then follows. As already mentioned this implies $\mathfrak{v}_\zeta - 1 \to 0$ in L_p $\forall p \in [2 - \varepsilon_0, 2 + \varepsilon_0]$ as $\lambda \to 0$. Finally, the first assertion in 1.18 follows from this in conjunction with 1.17 and theorem 0.4.3(I). The proof of 1.18 is complete.

By 1.12(I), 1.13 we have:

1.22 *$\mathfrak{F}(w, \lambda) := \mathfrak{v} \circ F(w, \lambda)$ is a hydrodynamically normalized $(\varkappa, -\varkappa)$-solution in $G_\mathfrak{c} \supset G_w$. (Note that also $\mathfrak{v}$ depends on λ.)*

From 1.18, 1.10 we now at once obtain

1.23 $\mathfrak{F}(w, \lambda) - w \to 0$ uniformly in every $G' \Subset G_\mathfrak{c}$ as $\lambda \to 0$,

and because of

$$\|\mathfrak{F}_w(w, \lambda) - 1\|_{L_p(G')} \leqq \|(\mathfrak{v}_\zeta - 1) \cdot F'(w, \lambda)\|_{L_p(G')} + \|F'(w, \lambda) - 1\|_{L_p(G')},$$

1.18, 1.19 and the boundedness of the $F'(w, \lambda)$ in $G' \Subset G_\mathfrak{c}$ we also obtain

1.24 $\|\mathfrak{F}_w(w, \lambda) - 1\|_{L_p(G')} \to 0$ for every $p \in [2 - \varepsilon_0, 2 + \varepsilon_0]$ and $G' \Subset G_\mathfrak{c}$, as $\lambda \to 0$.

Those $\mathfrak{F}(w, \lambda)$ which belong to a sequence of conformal boundary variations $F(w, \lambda)$ with respect to $\mathfrak{c}$ and w_0 according to 1.22 *shall be called $\varkappa$-conformal boundary variations with respect to $\mathfrak{c}$ and w_0.*

The relations 1.22, 1.17 imply

$$\begin{aligned}\mathfrak{F}(w, \lambda) &= F(w, \lambda) - \frac{1}{\pi} \iint \frac{2i\varkappa_\lambda(\zeta) \operatorname{Im} [\mathfrak{F}_w(w(\zeta), \lambda)\, w'(\zeta)]}{\zeta - F(w, \lambda)} \, d\sigma_\zeta \\ &\quad - \frac{1}{\pi} \iint \frac{\varkappa_\lambda(\zeta) (\overline{w'(\zeta)} - w'(\zeta))\, \mathfrak{F}_w(w(\zeta), \lambda)}{\zeta - F(w, \lambda)} \, d\sigma_\zeta \\ &= F(w, \lambda) - \frac{1}{\pi} \iint \frac{2i\varkappa_\lambda(\zeta)\, \overline{w'(\zeta)} \operatorname{Im} \mathfrak{F}_w(w(\zeta), \lambda)}{\zeta - F(w, \lambda)} \, d\sigma_\zeta .\end{aligned}$$

Putting $\zeta = F(t, \lambda)$ and making t the new integration variable we obtain

1.25 $$F(w, \lambda) = F(w, \lambda) - \frac{1}{\pi} \iint \frac{2i\varkappa(t)\, F'(t, \lambda) \operatorname{Im} (\mathfrak{F}_t(t, \lambda) - 1)}{F(t, \lambda) - F(w, \lambda)} \, d\sigma_t .$$

Using the Cauchy integral formula (note that the $o(\lambda)$ mentioned in 1.10(II) depends analytically on w in $G_\mathfrak{c}$) we see

1.26 $$F'(w, \lambda) = 1 - \frac{\lambda}{(w - w_0)^2} + o(\lambda),$$

$$F(w, \lambda) - F(t, \lambda) = (w - t) \left(1 - \frac{\lambda}{(t - w_0)(w - w_0)} + o(\lambda)\right),$$

where both the $o(\lambda)$ in 1.26 again satisfy

1.27 $o(\lambda)/\lambda \to 0$ locally uniformly in $G_\mathfrak{c}$ as $\lambda \to 0$.

Observing $\varkappa(w) = 0$ in $G_w \setminus f(D)$, $f(D) \Subset G_w = f(G)$ and 1.24, we now obtain

1.28 $$\mathfrak{F}(w, \lambda) = w + \frac{\lambda}{w - w_0} + P(2i\varkappa \operatorname{Im} (\mathfrak{F}_w - 1))(w) + o(\lambda),$$

where $o(\lambda)$ from 1.28 again has the property named in 1.27. Since $\mathfrak{F}(w, \lambda)$ is a $(\varkappa, -\varkappa)$-solution in $G_c \supset G_w$, by means of the differentiation rules for the P-operator (see 0.4.19) the relation

1.29
$$(o(\lambda))_{\bar{w}} = 0 \quad \text{in } G_c$$

follows. Hence, by Weyl's lemma 0.3.13, this $o(\lambda)$ is an analytic function of w in G_c. Because of the hydrodynamical normalization of $\mathfrak{F}(w, \lambda)$ and by 0.4.3(IV), this $o(\lambda)$ has a zero at infinity, that is

1.30
$$o(\lambda) = \frac{a_1(\lambda)}{w} + \frac{a_2(\lambda)}{w^2} + \ldots \quad \text{for } |w| \geqq R^*,$$

where

1.31
$$a_j(\lambda) = \frac{1}{2\pi i} \oint_{|w|=R^*} o(\lambda)\, w^{j-1}\, dw, \quad \text{thus } |a_j(\lambda)| \leqq R^{*j} \cdot \max\{|o(\lambda)| : |w| = R^*\}.$$

Differentiating both sides in 1.28 gives

1.32
$$\mathfrak{F}_w(w, \lambda) = 1 - \frac{\lambda}{(w - w_0)^2} + T(2i\varkappa \operatorname{Im}(\mathfrak{F}_w - 1))(w) + o^*(\lambda),$$

$o^*(\lambda) = (o(\lambda))_w$ with $o(\lambda)$ from 1.28. By the Cauchy integral formula we clearly again have

1.33
$$o^*(\lambda)/\lambda \to 0 \text{ locally uniformly in } G_c \text{ as } \lambda \to 0.$$

From this and from 1.30, 1.31 we also obtain

1.34 $\|o^*(\lambda)/\lambda\|_{L_p(G')} \to 0$ as $\lambda \to 0$, for every $G' \Subset G_c$ and every $p > 1$.

Now, $\mathfrak{F}$ occurs on both sides in 1.28 and 1.32. In order to remove this deficiency we again consider a sequence of conformal boundary variations $F(w, \lambda)$ with respect to $\mathfrak{c}$ and w_0, and let $\alpha_\lambda = \arg \lambda$, $0 \leqq \alpha_\lambda \leqq 2\pi$. After proceeding to a subsequence, if necessary, we can without loss of generality assume that $\lim_{\lambda \to 0} \alpha_\lambda =: \alpha$ exists. We now put

1.35
$$s(w, w_0, \alpha, \lambda) = (\mathfrak{F}_w(w, \lambda) - 1)/|\lambda|.$$

By 1.32 we then have

1.36
$$s(w, w_0, \alpha, \lambda) = -\frac{e^{i\alpha_\lambda}}{(w - w_0)^2} + \frac{o^*(\lambda)}{|\lambda|} + T(2i\varkappa \operatorname{Im} s(\cdot, w_0, \alpha, \lambda))(w).$$

By comparing the imaginary parts on both sides in 1.36, for $h := \operatorname{Im} s(w, w_0, \alpha, \lambda)$ the relation

1.37
$$h = g(\cdot, \lambda) + \mathfrak{T}h$$

follows, with

1.38
$$g(w, \lambda) := -\operatorname{Im}\left(\frac{e^{i\alpha_\lambda}}{(w - w_0)^2} - \frac{o^*(\lambda)}{|\lambda|}\right) = -\operatorname{Im}\left(\frac{e^{i\alpha}}{(w - w_0)^2} - \frac{o(\lambda)}{\lambda}\right)$$

and

1.39 $$\mathfrak{T}h := T(\varkappa h) + \overline{T(\varkappa \bar{h})} = -\frac{1}{\pi}\iint h(t)\,\mathrm{Re}\frac{2\varkappa(t)}{(t-w)^2}\,d\sigma_t.$$

For $o(\lambda)$ in 1.38 again 1.33 and 1.34 hold.

VI.2 THE VARIATIONAL FORMULAE FOR $\varkappa$-CONFORMAL BOUNDARY VARIATION

Next we have to consider the operator $\mathfrak{T}$ defined in 1.39 in more detail. In the variational formulae we are looking for, the operator

2.1 $$\tilde{\mathfrak{T}}h(w) := T(2i\varkappa\,\mathrm{Im}\,h)\,(w),$$

which is in fact non- (complex-) linear but at least additive, will also play a role. Thus we shall investigate both these operators simultaneously. We have

2.2 $$\mathrm{Im}\,(\tilde{\mathfrak{T}}h) = \mathfrak{T}(\mathrm{Im}\,h) = \mathrm{Im}\,(\mathfrak{T}h), \qquad \mathrm{Re}\,(\mathfrak{T}h) = \mathfrak{T}(\mathrm{Re}\,h).$$

Let $G^\varkappa$ be a measurable set such that $|\mathrm{supp}_0\,\varkappa \setminus G^\varkappa| = 0$, $\varkappa$ as in 1.5. For any $h \in L_p(G^\varkappa)$ we put $\varkappa(w)\,h(w) = 0$ if $w \notin G^\varkappa$. Thus, $\varkappa h \in L_p$ if $h \in L_p(G^\varkappa)$.

2.3 *$\mathfrak{T}$ and $\tilde{\mathfrak{T}}$ map $L_p(G^\varkappa)$ into L_p. They are contraction operators in $L_p(G^\varkappa)$ for every $p \in [2-\varepsilon_0, 2+\varepsilon_0]$ (cf. 1.16) and every $G^\varkappa$. Namely, for these p and $G^\varkappa$ it holds*

$$\|\mathfrak{T}\|_{L_p(G^\varkappa)} \leqq k', \qquad \|\tilde{\mathfrak{T}}\|_{L_p(G^\varkappa)} \leqq k'$$

(with k' from 1.16).

Indeed, for example, with $\mathfrak{T}$ we have

$$\|\mathfrak{T}h\|_{L_p(G^\varkappa)} \leqq \|\mathfrak{T}h\|_{L_p} = \|T(\varkappa h) + \overline{T(\varkappa\bar{h})}\|_{L_p} \leqq 2C(p)\,\|\varkappa h\|_{L_p}$$
$$\leqq 2C(p)\,\|\varkappa\|_{L_\infty}\,\|h\|_{L_p(G^\varkappa)} \leqq k'\|h\|_{L_p(G^\varkappa)}$$

for the named p and $G^\varkappa$. The assertion for $\tilde{\mathfrak{T}}$ follows analogously.

Because of the additivity of $\mathfrak{T}$ and $\tilde{\mathfrak{T}}$ and 2.3 it then holds:

2.4 *Let $p \in [2-\varepsilon_0, 2+\varepsilon_0]$. If $f_n \to f$ in $L_p(G^\varkappa)$ as $n \to \infty$ and if h_n is a solution of the equation*

(1) $h_n = f_n + \mathfrak{T}h_n$, *respectively* $h_n = f_n + \tilde{\mathfrak{T}}h_n$,

in $L_p(G^\varkappa)$, then $h_n \to h$ in $L_p(G^\varkappa)$ as $n \to \infty$, where h is the unique solution of the equation

(2) $h = f + \mathfrak{T}h$, *respectively* $h = f + \tilde{\mathfrak{T}}h$,

in $L_p(G^\varkappa)$.

We now turn back to equation 1.36. On account of 2.4 we obtain

2.5 $$s(w, w_0, \alpha, \lambda) = s(w, w_0, \alpha) + \frac{o(\lambda)}{\lambda}, \quad \frac{o(\lambda)}{\lambda} \to 0 \text{ in } L_p(G^\varkappa) \text{ as } \lambda \to 0,$$

where $s(w, w_0, \alpha)$ is the solution of the equation

2.6 $$s(w, w_0, \alpha) = -\frac{e^{i\alpha}}{(w - w_0)^2} + \tilde{\mathfrak{T}}s(\cdot, w_0, \alpha)(w)$$

in $L_p(G^\varkappa)$ with any $p \in [2 - \varepsilon_0, 2 + \varepsilon_0]$, but where $G^\varkappa$, apart from the condition $|\operatorname{supp}_0 \varkappa \setminus G^\varkappa| = 0$, must have positive distance from $w_0 \in \mathfrak{c} \subset \partial G_w$ $\left(\text{in order that } \frac{1}{(w - w_0)^2} \in L_p(G^\varkappa)\right)$.

When comparing the imaginary parts on both sides of 2.6 and using 2.2, we see that $h = \operatorname{Im} s(w, w_0, \alpha)$ satisfies

2.7 $$h(w) = -\operatorname{Im} \frac{e^{i\alpha}}{(w - w_0)^2} + \mathfrak{T}h(w) = \operatorname{Re} \frac{ie^{i\alpha}}{(w - w_0)^2} + \mathfrak{T}h(w).$$

Now let $\psi_\gamma(w, w_0)$ be the solution of

2.8 $$\psi_\gamma(w, w_0) = \frac{ie^{i\gamma}}{(w - w_0)^2} + \mathfrak{T}\psi_\gamma(\cdot, w_0)(w)$$

in $L_p(G^\varkappa)$. According to the second equation in 2.2 we then have

2.9 $$\operatorname{Im} s(w, w_0, \alpha) = \operatorname{Re} \psi_\alpha(w, w_0) = \operatorname{Re} [e^{i\alpha}\psi_0(w, w_0)],$$

the latter because $\mathfrak{T}$ is linear. For brevity we put

2.10 $$\psi(w, w_0) := \psi_0(w, w_0)$$

(i.e. we omit the index γ if $\gamma = 0$).

From 1.35, 2.5, 2.9, 2.10, 1.28, 0.4.3 it now follows

2.11 $$\mathfrak{F}(w, \lambda) = w + \frac{\lambda}{w - w_0} + P(2i\varkappa \operatorname{Re}(\lambda\psi(\cdot, w_0)))(w) + o(\lambda),$$

where this $o(\lambda)$ satisfies 1.27. In the same way, from 1.35, 2.5, 2.9, 2.10, 1.32 and from the continuity of $\tilde{\mathfrak{T}}$ the relation

2.12 $$\mathfrak{F}_w(w, \lambda) = 1 - \frac{\lambda}{(w - w_0)^2} + \tilde{\mathfrak{T}}(i\lambda\psi(\cdot, w_0))(w) + o(\lambda)$$

follows, where

2.13 $\|o(\lambda)/\lambda\|_{L_p(G')} \to 0$ for every $p \in [2 - \varepsilon_0, 2 + \varepsilon_0]$ and every $G' \Subset G_c$ as $\lambda \to 0$. In contrast to $o(\lambda)$ and $o^*(\lambda)$ from 1.28 and 1.32, respectively, the $o(\lambda)$ in 2.11 and 2.12 are not analytic functions of w in G_c throughout. Consequently, $o(\lambda)$ in 2.12 no longer satisfies 1.27. However, by means of the Cauchy integral formula, also for $o(\lambda)$ from 2.12 we, of course, obtain:

2.14 *If $\lambda \to 0$ then $o(\lambda)/\lambda \to 0$ locally uniformly in every open set $\subset G_c$ where $\varkappa(w) = 0$ a.e.*

We now form varied functions $f(\cdot, \lambda)$ from $f \in \mathfrak{Q}(\nu, \mu)$ by setting

2.15 $$f(z, \lambda) := \mathfrak{F}(f(z), \lambda) \in \mathfrak{Q}(\nu, \mu),$$

where $\mathfrak{F}(\cdot, \lambda)$ are $\varkappa$-conformal boundary variations with respect to $\mathfrak{c} \subset \partial G_w = \partial f(G)$ and $w_0 \in \mathfrak{c}$, and where $\varkappa = \varkappa_f$. The desired variational formulae for these $f(\cdot, \lambda)$ we will obtain from 2.11, 2.12. To start with, from 2.11, we immediately infer

2.16 $$f(z, \lambda) = f(z) + \frac{\lambda}{f(z) - w_0} - \frac{1}{\pi}\iint \frac{2i\varkappa(t)\operatorname{Re}(\lambda\psi(t, w_0))}{t - f(z)}\, d\sigma_t + o(\lambda),$$

where $o(\lambda)/\lambda \to 0$ locally uniformly in G as $\lambda \to 0$.

Further, when using $f_z(z, \lambda) = \mathfrak{F}_w(f(z), \lambda) f_z + \mathfrak{F}_{\bar{w}}(f(z), \lambda)\overline{f_{\bar{z}}}$, $\mathfrak{F}_{\bar{w}} = 2i\varkappa \operatorname{Im} \mathfrak{F}_w$ as well as

2.17

$$\operatorname{Im} \mathfrak{F}_w = -\operatorname{Im}\frac{\lambda}{(w - w_0)^2} - \operatorname{Im}(i\lambda\mathfrak{T}\psi(\cdot, w_0)) + o(\lambda) = \operatorname{Re}[\lambda\psi(\cdot, w_0)] + o(\lambda)$$

(which is a consequence of 2.12, 2.2, 2.8, 2.10), we finally obtain

2.18 $$f_z(z, \lambda) = f_z(z) - \frac{\lambda f_z(z)}{(f(z) - w_0)^2} + f_z\tilde{\mathfrak{T}}(i\lambda\psi(\cdot, w_0))(f(z)) + \overline{f_{\bar{z}}(z)} \cdot 2i\varkappa(f(z)) \operatorname{Re}(\lambda\psi(f(z), w_0)) + o(\lambda),$$

where $o(\lambda)$ in 2.18, as a function of z in G, satisfies

2.19 $$\|o(\lambda)/\lambda\|_{L_2(G')} \to 0 \text{ for every } G' \Subset G \text{ as } \lambda \to 0.$$

If $|\nu(z)| + \mu(z)| = 0$ a.e. in a neighbourhood of $z_1 \in G$ then $\varkappa(w) = \varkappa_f(w) = 0$ a.e. in a neighbourhood of $w_1 = f(z_1)$. Hence $o(\lambda)$ in 2.16 is analytic in a neighbourhood of z_1. By means of the Cauchy integral formula, for every natural number n, it then follows

2.20 $$f^{(n)}(z, \lambda) = f^{(n)}(z) + \lambda \frac{d^n \dfrac{1}{f(z) - w_0}}{dz^n} + \frac{1}{\pi}\iint 2i\varkappa(t)[\operatorname{Re}(\lambda\psi(t, w_0))] \frac{d^n \dfrac{1}{f(z) - t}}{dz^n}\, d\sigma_t + o(\lambda)$$

for all z from a neighbourhood of z_1, where $o(\lambda)/\lambda$ uniformly tends to zero in this neighbourhood of z_1 as $\lambda \to 0$.

The asymptotical expansions 2.16, 2.18, 2.20 are the variational formulae for $\varkappa$-conformal boundary variation we were looking for. Those $f(z, \lambda)$ which, according to 2.15, belong to a sequence of $\varkappa$-conformal boundary variations with respect to $\mathfrak{c}$ and w_0 are to be called $\varkappa$-conformal boundary variations *of f* with respect to $\mathfrak{c}$ and w_0.

VI.3 THE VARIATIONAL FORMULAE FOR INNER VARIATION

Again let f be any fixed mapping from $\mathfrak{Q}(\nu, \mu)$. Let $\gamma(w)$ be any function defined and measurable in $\mathbb{C}$ and satisfying $|\gamma(w)| \leqq 1$ $\forall w \in \mathbb{C}$, $\gamma(w) = 0$ if $w \in f(D)$, and ϑ be any positive real number $\leqq k$ (k from 1.1).

3.1 *There exists just one quasiconformal mapping $\mathfrak{F}^0(w, \vartheta\gamma)$ of $\mathbb{C}$ onto itself such that $\mathfrak{F}^0$ is a solution of*

(I) $\mathfrak{F}^0_{\bar{w}} = 2i\varkappa \operatorname{Im} \mathfrak{F}^0_w + \vartheta\gamma\mathfrak{F}^0_w$

(where $\varkappa = \varkappa_f$ belongs to f according to 1.5) *and admits an expansion*

(II) $\mathfrak{F}^0(w, \vartheta\gamma) = w + \varepsilon_{\mathfrak{F}^0}(w)$ *with* $\lim_{w\to\infty} \varepsilon_{\mathfrak{F}^0}(w) = 0$

(i.e. $\mathfrak{F}^0$ is hydrodynamically normalized with respect to the system (I)).

Proof. The uniqueness follows as with 1.12. In order to prove existence we consider, analogously with 1.14, the equation

3.2
$$h = T(\vartheta\gamma) + T(2i\varkappa \operatorname{Im} h + \vartheta\gamma h).$$

With $h \in L_p$ we have

3.3
$$\|T(2i\varkappa \operatorname{Im} h + \vartheta\gamma h)\|_{L_p} \leqq C(p)\,\|2|\varkappa h| + \vartheta|\gamma h|\|_{L_p}$$
$$\leqq C(p)\Big[\|2\varkappa\|^p_{L_\infty}\cdot \iint_{f(D)} |h|^p\, d\sigma_w + \vartheta^p \iint_{\mathbb{C}\setminus f(D)} |h|^p\, d\sigma_w\Big]^{1/p} \leqq C(p)\max\{\vartheta, 2\|\varkappa\|_{L_\infty}\}\,\|h\|_{L_p}$$
$$\leqq C(p)\left(1 - \frac{(1-k)^2}{1+k^2}\right)\|h\|_{L_p} \leqq k'\|h\|_{L_p} \quad \forall p \in [2-\varepsilon_0, 2+\varepsilon_0], \quad \forall \vartheta \in [0, k]$$

(cf. 1.16). By 1.4 supp $\gamma(\cdot)$ is always bounded, hence $\gamma \in L_p$ for every $p \geqq 1$. Thus 3.2 always has a unique solution in L_p for every $p \in [2-\varepsilon_0, 2+\varepsilon_0]$. Analogously with 1.17 we conclude that

3.4
$$\mathfrak{F}^0(w, \vartheta\gamma) := w + P(2i\varkappa \operatorname{Im} h + \vartheta\gamma(h+1))\,(w)$$

has all the properties mentioned in 3.1.

Consequently,

3.5
$$f^0(z, \vartheta\gamma) := \mathfrak{F}^0(f(z), \vartheta\gamma)$$

again belongs to $\mathfrak{Q}(\nu, \mu)$ as long as the corresponding variation of the dilatation quotient does not lead to a violation of the dilatation condition 1.3(1).

For variational purposes we again need an asymptotic expansion of $\mathfrak{F}^0(w, \vartheta\gamma)$ as $\vartheta \to 0$. This is our next aim.

For $\mathfrak{F}^0_w - 1 =: h$ we have, by 3.2 and 3.3,

3.6
$$\|\mathfrak{F}^0_w - 1\|_{L_p} \leqq \frac{\vartheta C(p)}{1-k'}\|\gamma\|_{L_p} \leqq M\vartheta \quad \text{for every } p, \vartheta \text{ as in 3.3,}$$

where the constant M depends only on p and on the area measure of the complement of $f(D)$. Hence (note 1.4), M can be chosen uniformly for every $f \in \mathfrak{Q}(\nu, \mu)$ and every $p \in [2-\varepsilon_0, 2+\varepsilon_0]$. From 3.6 and 3.2 we further infer

3.7
$$\mathfrak{F}^0_w - 1 = \vartheta T\gamma + T(2i\varkappa \operatorname{Im}(\mathfrak{F}^0_w - 1)) + O(\vartheta^2),$$

where

3.8
$$\|O(\vartheta^2)\|_{L_p} \leqq C(p)\,M\vartheta^2 \quad \forall p, \vartheta \text{ as in 3.3.}$$

With $s(w, \gamma, \vartheta) := (\mathfrak{F}_w^0 - 1)/\vartheta$ we then have

3.9 $$s(w, \gamma, \vartheta) = T\gamma + O(\vartheta) + \tilde{\mathfrak{T}}s(\cdot, \gamma, \vartheta), \quad \text{where } \|O(\vartheta)\|_{L_p} \leqq C(p)\, M\vartheta.$$

Thus, by 2.4 (cf. 2.5),

3.10 $$s(w, \gamma, \vartheta) = s(w, \gamma) + O^*(\vartheta) \quad \text{with } \|O^*(\vartheta)\|_{L_p} \leqq \frac{C(p)}{1 - k'} M\vartheta \quad \forall p, \vartheta \text{ as in 3.3},$$

where $s(w, \gamma)$ is the solution of

3.11 $$s(\cdot, \gamma) = T\gamma + \tilde{\mathfrak{T}}s(\cdot, \gamma).$$

Then (cf. 2.5–2.9) $\operatorname{Im} s(\cdot, \gamma) = \operatorname{Im} T\gamma + \mathfrak{T} \operatorname{Im} s(\cdot, \gamma)$, hence

3.12 $$\operatorname{Im} s(w, \gamma) = \operatorname{Im} i\psi^0(w, \gamma),$$

where $\psi^0(w, \gamma)$ is the unique solution of the equation

3.13 $$\psi^0(\cdot, \gamma) = -iT\gamma + \mathfrak{T}\psi^0(\cdot, \gamma)$$

in L_p $\forall p$ as in 3.3 (as to the connection between $\psi(w, w_0)$ and $\psi^0(w, \gamma)$, see 5.32 below). According to 3.7 and 3.10 we then have

3.14 $$\mathfrak{F}_w^0(w, \vartheta\gamma) = 1 + \vartheta[T\gamma(w) + \tilde{\mathfrak{T}}(i\psi^0(\cdot, \gamma))(w)] + O(\vartheta^2)$$

with

$$\|O(\vartheta^2)\|_{L_p} \leqq \frac{C(p)}{1 - k'} M\vartheta^2$$

for every $p \in [2 - \varepsilon_0, 2 + \varepsilon_0]$ and every $\vartheta \in [0, k]$. Finally, 3.4, 3.10, 3.12 imply

3.15 $$\mathfrak{F}^0(w, \vartheta\gamma) = w - \frac{\vartheta}{\pi} \iint \frac{\gamma(t) + 2i\varkappa(t) \operatorname{Re} \psi^0(t, \gamma)}{t - w} d\sigma_t + O(\vartheta^2),$$

where $|O(\vartheta^2)| \leqq |P(\vartheta\gamma h + 2i\varkappa \operatorname{Im} [\vartheta O^*(\vartheta)])|$, $O^*(\vartheta)$ from 3.10. On account of 0.4.3 (I), 3.6, and 3.10, $O(\vartheta^2)$ from 3.15 satisfies

3.16 $$|O(\vartheta^2)| \leqq K_{2+\varepsilon_0, 2-\varepsilon_0} \cdot \left[M\vartheta^2 + \frac{M}{1 - k'} \vartheta^2 \cdot 2\|\varkappa\|_{L_\infty} \cdot \max\{C(p) : |2 - p| \leqq \varepsilon_0\}\right]$$
$$\leqq K_{2+\varepsilon_0, 2-\varepsilon_0} \frac{M}{1 - k'} \vartheta^2 \quad \forall w \in \mathbb{C} \text{ and } \forall \vartheta \in [0, k].$$

Hence, for the corresponding $f^0(z, \vartheta\gamma)$ we obtain

3.17 $$f^0(z, \vartheta\gamma) = f(z) - \frac{\vartheta}{\pi} \iint \frac{\gamma(t) + 2i\varkappa(t) \operatorname{Re} \psi^0(t, \gamma)}{t - f(z)} d\sigma_t + O(\vartheta^2),$$

$O(\vartheta^2)$ as in 3.16, for every $z \in G$ and every $\vartheta \in [0, k]$.

Analogously with 2.18, from 3.14 the relation

3.18 $$f_z^0(z, \vartheta\gamma) = f_z(z) - \frac{\vartheta f_z(z)}{\pi} \iint \frac{\gamma(t) + 2i\varkappa(t) \operatorname{Re} \psi^0(t, \gamma)}{(t - f(z))^2} d\sigma_t$$
$$+ \vartheta[\gamma(f(z)) + 2i\varkappa(f(z)) \operatorname{Re} \psi^0(f(z), \gamma)]\overline{f_{\bar{z}}(z)} + O(\vartheta^2)$$

follows, where $\|O(\vartheta^2)/\vartheta^2\|_{L_2(G)}$ is uniformly bounded for every $\vartheta \in (0, k]$. Of course, a statement analogous with 2.20 also holds true.

The relations 3.14, 3.15, 3.17, 3.18 are the variational formulae for inner variation in $\mathfrak{Q}(\nu, \mu)$ we were looking for.

3.19 ***Remark.*** Inner variations in $\mathfrak{Q}(\nu, \mu)$ are possible whenever the set $\text{supp}_0\,(|\nu(z)| + |\mu(z)|) \setminus D$ has positive measure, no matter whether G has a non-degenerate boundary component or not. Moreover, by inner variation we can vary every $f \in \mathfrak{Q}(\nu, \mu)$ for which $\mathbb{C} \setminus f(G)$ has positive measure.

VI.4 TWO EXAMPLES OF FUNCTIONALS

In order to illustrate and motivate the following reasonings we shall consider two extremal problems. For the sake of simplicity, throughout this section we suppose $\nu(z) = \mu(z) = 0$ in a neighbourhood of infinity, say for $|z| \geqq R_0$. Then every $f \in \mathfrak{Q}(\nu, \mu)$ admits a Laurent expansion

4.1
$$f(z) = z + \frac{a_1}{z} + \frac{a_2}{z^2} + \ldots \quad \text{for } |z| > R_0.$$

Let the first example be the functional $\chi_1[f] = \text{Re}\, a_1$. Of course, this $\chi_1[f]$ is continuous with respect to convergence in $\mathfrak{Q}(\nu, \mu)$. Hence at least one extremal function of the extremal problem

4.2
$$\chi_1[f] \to \max \quad \text{for } f \in \mathfrak{Q}(\nu, \mu)$$

does exist. When applying the variational formula for $\varkappa$-conformal boundary variation (in case G has a non-degenerate boundary component) to any $f \in \mathfrak{Q}(\nu, \mu)$, we obtain an asymptotic expansion of the functional,

4.3
$$\begin{aligned}\chi_1[f(\cdot, \lambda)] &= \chi_1[f] + \text{Re}\, \lambda + \text{Re}\, \frac{1}{\pi} \iint 2i\varkappa(t)\, \text{Re}\, (\lambda \psi(t, w_0))\, d\sigma_t + o(\lambda) \\ &= \chi_1[f] + \text{Re} \left\{ \lambda \left(1 - \frac{1}{\pi} \iint \psi(t, w_0)\, \text{Im}\, [2\varkappa(t) \cdot 1]\, d\sigma_t \right) \right\} + o(\lambda)\end{aligned}$$

with $\varkappa(t) = \varkappa_f(t)$. By applying the variational formula for inner variation we obtain

4.4
$$\begin{aligned}\chi_1[f^0(\cdot, \vartheta\gamma)] &= \chi_1[f] + \text{Re}\, \frac{\vartheta}{\pi} \iint [\gamma(t) + 2i\varkappa(t)\, \text{Re}\, \psi^0(t, \gamma)]\, d\sigma_t + O(\vartheta^2) \\ &= \chi_1[f] + \frac{\vartheta}{\pi}\, \text{Re} \iint (\gamma(t) - \psi^0(t, \gamma)\, \text{Im}\, [2\varkappa(t) \cdot 1])\, d\sigma_t + O(\vartheta^2).\end{aligned}$$

If f is an extremal function of χ_1, then from 4.3 the inequality

4.5
$$\text{Re} \left\{ e^{i\alpha} \left(1 - \frac{1}{\pi} \iint \psi(t, w_0)\, \text{Im}\, (2\varkappa(t) \cdot 1)\, d\sigma_t \right) \right\} \leqq 0$$

follows for every α for which a sequence of conformal boundary variations $F(w, \lambda_n)$ of $f(G)$ exists such that $\lim\limits_{n \to \infty} \dfrac{\lambda_n}{|\lambda_n|} = e^{i\alpha}$. Analogously, from 4.4 we infer

4.6 $$\operatorname{Re} \iint (\gamma(t) - \psi^0(t, \gamma) \operatorname{Im}(2\varkappa(t) \cdot 1))\, d\sigma_t \leqq 0$$

for every $\gamma(\cdot)$ such that $\mathfrak{F}^0(f(z), \vartheta\gamma) \in \mathfrak{Q}(\nu, \mu)$ for every ϑ with $0 \leqq \vartheta < \vartheta_0$, where ϑ_0 is a certain positive constant.

As the second example we choose $\chi_2[f] = \operatorname{Re} f(z_0)$, where z_0 is to be a fixed point from G. This $\chi_2[f]$ is continuous again, hence there exists at least one extremal function of the extremal problem 4.2 with χ_1 replaced by χ_2. When applying the boundary variational formula (if there is a non-degenerate boundary component) to any $f \in \mathfrak{Q}(\nu, \mu)$ we obtain

4.3′

$$\chi_2[f(\cdot, \lambda)] = \chi_2[f] + \operatorname{Re}\left(\frac{\lambda}{f(z_0) - w_0} - \frac{1}{\pi}\iint \frac{2i\varkappa(t) \operatorname{Re}(\lambda\psi(t, w_0))}{t - f(z_0)}\, d\sigma_t\right) + o(\lambda)$$

$$= \chi_2[f] + \operatorname{Re}\left\{\lambda\left(\frac{1}{f(z_0) - w_0} + \frac{1}{\pi}\iint \psi(t, w_0) \operatorname{Re}\frac{2i\varkappa(t)}{f(z_0) - t}\, d\sigma_t\right)\right\} + o(\lambda).$$

The variational formula 3.17 for inner variation yields

4.4′

$$\chi_2[f^0(\cdot, \vartheta\gamma)] = \chi_2[f] + \operatorname{Re}\left(\frac{\vartheta}{\pi}\iint \frac{\gamma(t) + 2i\varkappa(t) \operatorname{Re}\psi^0(t, \gamma)}{f(z_0) - t}\, d\sigma_t\right) + O(\vartheta^2)$$

$$= \chi_2[f] + \frac{\vartheta}{\pi}\operatorname{Re}\iint\left(\frac{\gamma(t)}{f(z_0) - t} - \psi^0(t, \gamma) \operatorname{Im}\frac{2\varkappa(t)}{f(z_0) - t}\right) d\sigma_t + O(\vartheta^2).$$

Analogously with 4.5 and 4.6, for extremal f we now obtain

4.5′ $$\operatorname{Re}\left\{e^{i\alpha}\left(\frac{1}{f(z_0) - w_0} - \frac{1}{\pi}\iint \psi(t, w_0) \operatorname{Im}\frac{2\varkappa(t)}{f(z_0) - t}\, d\sigma_t\right)\right\} \leqq 0$$

and

4.6′ $$\operatorname{Re}\iint\left(\frac{\gamma(t)}{f(z_0) - t} - \psi^0(t, \gamma) \operatorname{Im}\frac{2\varkappa(t)}{f(z_0) - t}\right) d\sigma_t \leqq 0$$

for the corresponding α and $\gamma(\cdot)$.

The inequalities 4.5, 4.5′ show the necessity to obtain insight in the nature of expressions of the form

4.7 $$A(w) - \frac{1}{\pi}\iint \psi(t, w) \operatorname{Im}(2\varkappa(t)\, A(t))\, d\sigma_t,$$

where $A(w)$ are certain analytic functions. This is our aim in the next section. In this connection the expressions of the form 4.6, 4.6′ will prove to be a kind of continuous analogue of 4.7.

VI.5 SOME TRANSFORMATIONS

Let Ω be a domain $\subseteq \mathbb{C}$ and $\varkappa$ be any function defined and measurable in $\mathbb{C}$ such that

5.1 $$2\|\varkappa\|_{L_\infty} \leqq 1 - \frac{(1 - k)^2}{1 + k^2} =: 2k_0 < 1, \quad |\operatorname{supp}_0 \varkappa \setminus \Omega| = 0.$$

With this $\varkappa$ let $\mathfrak{T}$ be defined as in 1.39, and further let (cf. 0.4.29)

5.2 $$\mathfrak{T}_\Omega^* h(w) := \varkappa(w)\, T_\Omega h(w) + \overline{\varkappa(w)}\, \overline{T_\Omega \bar{h}(w)} = -\frac{1}{\pi} \iint_\Omega h(t) \operatorname{Re} \frac{2\varkappa(w)}{(t-w)^2}\, d\sigma_t$$

as $h \in L_{p,q}(\Omega)$. Analogously with 2.3 we then have

5.3 *$\mathfrak{T}$ and $\mathfrak{T}_\Omega^*$ map $L_p(\Omega)$ into L_p, $L_{p,q}(\Omega)$ into $L_{p,q}$ and $L^{p,q}(\Omega)$ into $L^{p,q}$, for every finite $p > 1$, $q > 1$. If $p, q \in [2 - \varepsilon_0, 2 + \varepsilon_0]$ (cf. 1.16) $\mathfrak{T}$ and $\mathfrak{T}_\Omega^*$ are contraction operators in $L_p(\Omega)$, $L_{p,q}(\Omega)$, and $L^{p,q}(\Omega)$, and it holds*

$$\|\mathfrak{T}\|_{\mathscr{H}} \leqq k', \qquad \|\mathfrak{T}_\Omega^*\|_{\mathscr{H}} \leqq k'$$

where $\mathscr{H}$ means any of the spaces $L_p(\Omega)$, $L_{p,q}(\Omega)$ and $L^{p,q}(\Omega)$.

Indeed, with the named p, q and, say, $h \in L_{p,q}(\Omega)$, we have

$$\begin{aligned}
\|\mathfrak{T}_\Omega^* h\|_{L_{p,q}(\Omega)} &\leqq \|\mathfrak{T}_\Omega^* h\|_{L_{p,q}} \leqq k_0(\|T_\Omega h\|_{L_{p,q}} + \|T_\Omega \bar{h}\|_{L_{p,q}}) \\
&\leqq k_0 \inf\{\|T_\Omega h_1\|_{L_q} + \|T_\Omega h_2\|_{L_p} : h = h_1 + h_2,\ h_1 \in L_q(\Omega), \\
&\qquad h_2 \in L_p(\Omega)\} + k_0\|T_\Omega \bar{h}\|_{L_{p,q}} \\
&\leqq k_0 \max\{C(p), C(q)\} \inf\{\|h_1\|_{L_q(\Omega)} + \|h_2\|_{L_p(\Omega)} : h = h_1 + h_2, \\
&\qquad h_1 \in L_q(\Omega), h_2 \in L_p(\Omega)\} + k_0\|T_\Omega \bar{h}\|_{L_{p,q}} \\
&\leqq \frac{k'}{2}\|h\|_{L_{p,q}(\Omega)} + k_0\|T_\Omega \bar{h}\|_{L_{p,q}} \leqq k'\|h\|_{L_{p,q}(\Omega)},
\end{aligned}$$

according to the definition of the norm in $L_{p,q}(\Omega)$ (cf. 0.4.22) and because $\|T_\Omega \bar{h}\|_{L_{p,q}}$, of course, satisfies the same estimate as $\|T_\Omega h\|_{L_{p,q}}$ does.

Let p, q be, from now on, any fixed numbers satisfying

5.4 $$p \in (2, 2 + \varepsilon_0], \qquad q \in [2 - \varepsilon_0, 2).$$

Moreover, let p, q be chosen in such a way that for p', q' satisfying $p^{-1} + p'^{-1} = 1$, $q^{-1} + q'^{-1} = 1$ it also holds

5.5 $$p' \in [2 - \varepsilon_0, 2), \qquad q' \in (2, 2 + \varepsilon_0].$$

Let $\phi(t, w)$ be the solution of

5.6 $$\phi(t, w) = \frac{\mathrm{i}}{t - w} + \mathfrak{T}\phi(\cdot, w)(t)$$

in $L_{p',q}$, where w is to be arbitrary but fixed $\in \mathbb{C}$. Then the operator

5.7 $$\Pi_\Omega f(w) := \frac{\mathrm{i}}{\pi} \iint_\Omega \phi(t, w) f(t)\, d\sigma_t$$

is in any case defined on $L^{p,q}(\Omega)$. Further, let

5.8 $$f^*(t) := f(t) - \mathfrak{T}_\Omega^* f(t),$$

where f belongs to one of the function spaces named in 5.3. If $f \in L^{p,q}(\Omega)$ then also $f^* \in L^{p,q}(\Omega)$, and we have

5.9 $$\Pi_\Omega f^*(w) = \frac{i}{\pi}\iint\limits_\Omega \phi(t, w) f^*(t)\, d\sigma_t = \frac{i}{\pi}\iint\limits_\Omega \phi(t, w)\,(f(t) - \mathfrak{T}^*_\Omega f(t))\, d\sigma_t$$
$$= \frac{i}{\pi}\iint\limits_\Omega f(t)\,[\phi(t, w) - \mathfrak{T}\phi(\cdot, w)(t)]\, d\sigma_t = P_\Omega f(w),$$

because $\mathfrak{T}^*_\Omega$ is the adjoint operator of $\mathfrak{T}$ in every space $\mathscr{H}$ mentioned in 5.3. That is, by 0.4.44 for $f \in \mathscr{H}$, $h \in \mathscr{H}^*$ (where $\mathscr{H}^*$ denotes the dual space of $\mathscr{H}$, cf. theorem 0.4.23) it holds

5.10 $$\iint\limits_\Omega f \cdot \mathfrak{T}^*_\Omega h\, d\sigma = \iint\limits_\Omega (\mathfrak{T}f)\, h\, d\sigma.$$

With $f^* \in L_{p,q}(\Omega)$ we now put

5.11 $$\Pi^*_\Omega f^*(w) := P^*_\Omega f(w).$$

Then for every $f \in L_{p,q}(\Omega)$, $\Pi^*_\Omega f$ represents a set of functions differing from each other only by additive constants (see 0.4.40). Further let

5.12 $$\theta_\Omega f^*(w) := T_\Omega f(w) \quad \text{for } f^* \in L_{p,q}(\Omega),$$

where f and f^* are to be connected here and in what follows by 5.8 (of course, for any prescribed $f^* \in \mathscr{H}$, with $\mathscr{H}$ as in 5.3 under the condition 5.4, equation 5.8 always has a unique solution f).

When using an analogous terminology as in 0.4.42, we then have, by 0.4.42,

5.13 $$(\Pi^*_\Omega f^*(w))_w = \theta_\Omega f^*(w), \qquad (\Pi^*_\Omega f^*(w))_{\bar w} = f(w) \cdot \omega_\Omega(w) \quad \text{a.e. in } \mathbb{C}$$

for every $f \in L_{p,q}(\Omega)$.

5.14 ***Definition.*** (A) Let $W^1_{p,q}(\Omega, \varkappa)$ be the set of all the functions satisfying

(I) $f \in L_{s,\mathrm{loc}}(\Omega)$, $s = q/(2 - q)$,

(II) f has generalized derivatives $f_w, f_{\bar w} \in L_{q,\mathrm{loc}}(\Omega)$ in Ω,

(III) $\mathrm{Im}\,(\varkappa f_w) \in L_{p,q}$ (where we put $\varkappa(w) f_w(w) = 0$ if $w \notin \Omega$).

(B) Let $[W^1_{p,q}(\Omega, \varkappa)]$ be the factor space of the real-linear space $W^1_{p,q}(\Omega, \varkappa)$ by the subspace of the functions constant a.e. in Ω, i.e. for $[g], [f] \in [W^1_{p,q}(\Omega, \varkappa)]$ $[f] = [g]$ if and only if $f - g =$ constant a.e. in Ω for every $f \in [f]$ and every $g \in [g]$.

For the transformation

5.15 $$\mathfrak{R}^*_\Omega[f] := [f] + i\Pi^*_\Omega(2\,\mathrm{Im}\,(\varkappa f_w)), \qquad [f] \in [W^1_{p,q}(\Omega, \varkappa)],\ f \in [f],$$

where $\Pi^*_\Omega(2\,\mathrm{Im}\,(\varkappa f_w))$ (like $P^*_\Omega(2\,\mathrm{Im}\,(\varkappa g_w))$, too, in 5.16 below) is now understood to be an element of $[W^1_{p,q}(\Omega, \varkappa)]$, we have the following.

5.16 Theorem. *$\mathfrak{R}^*_\Omega$ maps $[W^1_{p,q}(\Omega, \varkappa)]$ one-to-one onto itself, $\mathfrak{R}^*_\Omega[0] = [0]$. The inverse mapping $\mathfrak{R}^{*-1}_\Omega$ is given by*

$$\mathfrak{R}^{*-1}_\Omega[g] = [g] - iP^*_\Omega(2\,\mathrm{Im}\,(\varkappa g_w)).$$

Proof. First we have to show that $\mathfrak{R}^*_\Omega$ and $\mathfrak{R}^{*-1}_\Omega$ map $[W^1_{p,q}(\Omega, \varkappa)$ into itself. We have

5.17 $$P^*_\Omega h \in [W^1_{p,q}(\mathbb{C}, \varkappa)] \subseteq [W^1_{p,q}(\Omega, \varkappa)] \quad \text{for } h \in L_{p,q}.$$

Indeed, let $h = h_1 + h_2$ be any decomposition of h where $h_1 \in L_q$, $h_2 \in L_p$. Then clearly $P_{0\Omega}h_2 \in W^1_{p,q}(\mathbb{C}, \varkappa)$, cf. 0.4.3, 0.4.19. As to $P_\Omega h_1$, (II) and (III) of 5.14 are of course satisfied, by 0.4.42. Now let Ω' be any bounded domain and choose R to be large, so that $\Omega' \subseteqq \{|w| < R/2\}$. In the expression

$$P_\Omega h_1(w) = -\frac{1}{\pi} \iint\limits_{\Omega \cap \{|t| < R\}} \frac{h_1(t)}{t - w}\, d\sigma_t - \frac{1}{\pi} \iint\limits_{\Omega \cap \{|t| > R\}} \frac{h_1(t)}{t - w}\, d\sigma_t$$

(cf. 0.4.38) the second integral on the right-hand side is an analytic function of w in $\overline{\Omega}'$, and the first integral belongs to $L_s(\Omega')$ for every s subject to $1 < s < 2q/(2 - q)$, especially for $s = q/(2 - q)$, by theorem 0.4.31. Hence $P_\Omega h_1$ also satisfies (I) in 5.14. This proves 5.17. Consequently, $\mathfrak{K}^{*-1}_\Omega$ maps $[W^1_{p,q}(\Omega, \varkappa)]$ into itself. In order to prove the same for $\mathfrak{K}^*_\Omega$ it is enough now to note that, for every $h \in L_{p,q}$, the h_* defined by $(h_*)^* = h = h_* - \mathfrak{T}^*_\Omega h_*$ also belongs to $L_{p,q}$, and that $\Pi^*_\Omega h = P^*_\Omega h_*$.

We still have to show

5.18 $\quad \mathfrak{K}^*_\Omega(\mathfrak{K}^{*-1}_\Omega[g]) = [g]$ and $\mathfrak{K}^{*-1}_\Omega(\mathfrak{K}^*_\Omega[f]) = [f] \quad \forall f, g \in W^1_{p,q}(\Omega, \varkappa)$.

For example, in case of the second relation in 5.18 we conclude as follows: For any $h \in L_{p,q}$ henceforth let h_* denote that function from $L_{p,q}$ which satisfies $(h_*)^* = h$. Then we have

$$\mathfrak{K}^{*-1}_\Omega(\mathfrak{K}^*_\Omega[f]) - [f] = \mathrm{i}\Pi^*_\Omega(2\,\mathrm{Im}\,(\varkappa f_w)) - \mathrm{i}P^*_\Omega(2\,\mathrm{Im}\,\{\varkappa(f_w + \mathrm{i}\theta_\Omega(2\,\mathrm{Im}\,(\varkappa f_w)))\})$$

(note the first relation in 5.13). According to 5.11 this is

$$= \mathrm{i}P^*_\Omega\{(2\,\mathrm{Im}\,(\varkappa f_w))_* - 2\,\mathrm{Im}\,(\varkappa f_w) - 2\,\mathrm{Im}\,\{\mathrm{i}\varkappa T_\Omega(2\,\mathrm{Im}\,(\varkappa f_w))_*\}\} = \mathrm{i}P^*_\Omega(0) = [0],$$

by

5.19 $\quad 2\,\mathrm{Re}\,(\varkappa(t)\,T_\Omega h(t)) = \mathfrak{T}^*_\Omega h(t)$ for every real-valued $h \in L_{p,q}(\Omega)$

(note that if h is real-valued, so is h_*) and by 5.8. Quite analogously we obtain the first relation in 5.18. Since $\mathfrak{K}^*_\Omega$ and $\mathfrak{K}^{*-1}_\Omega$ are real-linear operators we of course have $\mathfrak{K}^*_\Omega[0] = [0]$. Theorem 5.16 is proved.

5.20 Theorem. *Let* $B \in W^1_{p,q}(\Omega, \varkappa)$ *be analytic in* $\Omega' \subseteqq \Omega$. *Then every* $\mathfrak{B} \in \mathfrak{K}^*_\Omega[B]$, *possibly after alteration on a null set, is a solution of the system*

(I) $\mathfrak{B}_{\overline{w}} = 2\mathrm{i}\,\mathrm{Im}\,(\varkappa\mathfrak{B}_w)$ *in* Ω'.

Proof. Because $B \in \mathfrak{K}^{*-1}_\Omega[\mathfrak{B}]$ we have, according to 0.4.42, $0 = B_{\overline{w}} = (\mathfrak{K}^{*-1}_\Omega[\mathfrak{B}])_{\overline{w}} = \mathfrak{B}_{\overline{w}} - \mathrm{i}(P^*_\Omega(2\,\mathrm{Im}\,(\varkappa\mathfrak{B}_w)))_{\overline{w}} = \mathfrak{B}_{\overline{w}} - 2\mathrm{i}\,\mathrm{Im}\,(\varkappa\mathfrak{B}_w)$ a.e. in Ω'. Now let $W(w)$ be a schlicht solution of the Beltrami system

5.21 $$W_{\overline{w}} = \lambda W_w, \qquad \lambda(w) = \begin{cases} \varkappa(w) - \overline{\varkappa(w)}\,\overline{(\mathfrak{B}_w/\mathfrak{B}_w)} & \text{if } w \in \Omega,\ B_w(w) \neq 0,\\ 0 & \text{otherwise.}\end{cases}$$

Then lemma I.1.7(II) (note 5.5 and 1.16) gives that $\mathfrak{B}$ coincides with an $L_{q'}$-solution of the $(\varkappa, -\overline{\varkappa})$-system in Ω', up to a null set. This proves 5.20.

Let

5.22 $$L_{p,q}(\Omega, \varkappa) = \{f : f \in L_{q,\mathrm{loc}}(\Omega),\ \mathrm{Im}\,(\varkappa f) \in L_{p,q}\}$$

(again $\varkappa f = 0$ outside Ω). For the transformation

5.23 $$\mathfrak{L}_\Omega f := f + \mathrm{i}\theta_\Omega(2\,\mathrm{Im}\,(\varkappa f)), \quad f \in L_{p,q}(\Omega, \varkappa)$$

the relation

5.24 $$(\mathfrak{K}_\Omega^*[f])_w = \mathrm{L}_\Omega(f_w) \quad \text{for every } f \in W^1_{p,q}(\Omega, \varkappa)$$

holds, by 5.13.

5.25 Theorem. *$\mathfrak{L}_\Omega$ is a one-to-one mapping of $L_{p,q}(\Omega, \varkappa)$ onto itself. The inverse mapping of $\mathfrak{L}_\Omega$ is*

$$\mathfrak{L}_\Omega^{-1} g := g - \mathrm{i}T_\Omega(2\,\mathrm{Im}\,(\varkappa g)).$$

Proof. By the properties of the T- and the θ_Ω-operator it is clear that $\mathfrak{L}_\Omega$ and $\mathfrak{L}_\Omega^{-1}$ (as defined in 5.25) map $L_{p,q}(\Omega, \varkappa)$ into itself. We still have to show

5.26 $$\mathfrak{L}_\Omega^{-1}(\mathfrak{L}_\Omega f) = f \quad \text{and} \quad \mathfrak{L}_\Omega(\mathfrak{L}_\Omega^{-1} g) = g \quad \text{for every } f, g \in L_{p,q}(\Omega, \varkappa).$$

For instance, we have

$$\begin{aligned}\mathfrak{L}_\Omega^{-1}(\mathfrak{L}_\Omega f) &= f + \mathrm{i}\theta_\Omega(2\,\mathrm{Im}\,(\varkappa f)) - \mathrm{i}T_\Omega(2\,\mathrm{Im}\,\{\varkappa(f + \mathrm{i}\theta_\Omega(2\,\mathrm{Im}\,(\varkappa f)))\}) \\ &= f + \mathrm{i}T_\Omega\{(2\,\mathrm{Im}\,(\varkappa f))_* - 2\,\mathrm{Im}\,(\varkappa f) - 2\,\mathrm{Re}\,(\varkappa T_\Omega(2\,\mathrm{Im}\,(\varkappa f))_*)\} = f,\end{aligned}$$

by 5.19, 5.8. In the same way the second relation in 5.26 follows. Theorem 5.25 is proved.

The next theorem relates the expressions from 4.7 with the transformation $\mathfrak{L}_\Omega$.

5.27 Theorem. *For every $w \in \Omega \setminus \operatorname{supp} \varkappa$ and every $h \in L_{p,q}(\Omega, \varkappa)$*

$$\mathfrak{L}_\Omega h(w) = h(w) - \frac{1}{\pi} \iint \psi(t, w) \cdot 2\,\mathrm{Im}\,(\varkappa(t)\, h(t))\, \mathrm{d}\sigma_t$$

with $\psi(t, w)$ as in 2.10.

Proof. To prove theorem 5.27 we have to show

$$\mathrm{i}\theta_\Omega(2\,\mathrm{Im}\,(\varkappa h)) = -\frac{1}{\pi} \iint \psi(t, w) \cdot 2\,\mathrm{Im}\,(\varkappa(t)\, h(t))\, \mathrm{d}\sigma_t\, \forall w \notin \operatorname{supp} \varkappa, \forall h \in L_{p,q}(\Omega, \varkappa).$$

To this end we look for a kernel $K(t, w)$ which, for each fixed $w \notin \operatorname{supp} \varkappa$, is a function of t belonging to $L^{p',q'}(\operatorname{supp} \varkappa)$, where p' and q' are as in 5.5, so that it holds

5.28 $$\theta_\Omega(g \cdot \omega_{\operatorname{supp} \varkappa})^* = \frac{1}{\pi} \iint_{\operatorname{supp} \varkappa} K(t, w)\, (g \cdot \omega_{\operatorname{supp} \varkappa})^*\, \mathrm{d}\sigma_t \quad \forall w \notin \operatorname{supp} \varkappa$$

and for every $g \in L_{p,q}(\operatorname{supp} \varkappa)$ ($\omega_{\operatorname{supp} \varkappa}$ is to be the characteristic function of $\operatorname{supp} \varkappa$). As a necessary and sufficient condition for this we obtain, using 0.4.44,

$$\begin{aligned} 0 &= T_{\operatorname{supp} \varkappa} g(w) - \frac{1}{\pi} \iint_{\operatorname{supp} \varkappa} K(t, w) \left[g(t) + \frac{1}{\pi} \iint_{\operatorname{supp} \varkappa} g(s)\, \mathrm{Re}\, \frac{2\varkappa(t)}{(t - s)^2}\, \mathrm{d}\sigma_s \right] \mathrm{d}\sigma_t \\ &= -\frac{1}{\pi} \iint_{\operatorname{supp} \varkappa} g(t) \left[\frac{1}{(t - w)^2} + K(t, w) + \frac{1}{\pi} \iint K(s, w) \cdot \mathrm{Re}\, \frac{2\varkappa(s)}{(s - t)^2}\, \mathrm{d}\sigma_s \right] \mathrm{d}\sigma_t \end{aligned}$$

for every $g \in L_{p,q}(\operatorname{supp} \varkappa)$; here $g \cdot \omega_{\operatorname{supp} \varkappa}$ and $(g \cdot \omega_{\operatorname{supp} \varkappa})^*$ are again connected by 5.8. Hence $K(t, w)$ must be a solution of

5.29 $$K(t, w) = -\frac{1}{(t-w)^2} + \mathfrak{T}K(\cdot, w)(t).$$

Conversely, for every $w \notin \operatorname{supp} \varkappa$, 5.29 has a unique solution in $L^{p',q'}(\operatorname{supp} \varkappa)$, by 2.3. Hence this solution possesses all the properties needed for the operations carried out above. Using the linearity and unique solvability of 5.29, as well as observing 2.8 and 2.10, we obtain

5.30 $$K(t, w) = \mathrm{i}\psi(t, w),$$

which proves theorem 5.27.

Finally, let us draw the connection between $\psi^0(w, \gamma)$ and $\psi(t, w)$ as well as θ_Ω. By 3.13 and 2.3, $\psi^0(w, \gamma) \in L^{p',q'}$ for every γ mentioned at the beginning of VI.3. We have

5.31 $$\mathrm{i} \iint_\Omega \psi^0(w, \gamma) f(w)\, d\sigma_w = \iint \gamma(w)\, \theta_\Omega f(w)\, d\sigma_w \quad \forall f \in L_{p,q}(\Omega)$$

(as to the range of integration on the right-hand side of 5.31 let us recall the agreement near 0.3.1).

Proof. Let f^* be any function from $L_{p,q}(\Omega)$ and let f be the corresponding solution of 5.8 in $L_{p,q}(\Omega)$. Again by 0.4.44, respectively by 5.10, we have

$$\iint_\Omega \psi^0(w, \gamma) f^*(w)\, d\sigma_w = \iint_\Omega \psi^0(w, \gamma)\,(f(w) - \mathfrak{T}_\Omega^* f(w))\, d\sigma_w$$
$$= \iint_\Omega f(w)\,(\psi^0(w, \gamma) - \mathfrak{T}\psi^0(\cdot, \gamma)(w))\, d\sigma_w = \iint_\Omega f(w)\,(-\mathrm{i})\, T\gamma(w)\, d\sigma_w$$
$$= -\mathrm{i} \iint \gamma(w)\, T_\Omega f(w)\, d\sigma_w = -\mathrm{i} \iint \gamma(w)\, \theta_\Omega f^*(w)\, d\sigma_w,$$

which proves 5.31.

The relations 5.28, 5.30, 5.31 now immediately yield:

5.32 *If* $\operatorname{supp} \gamma \cap \operatorname{supp} \varkappa = \emptyset$ *then for every* $f \in L_{p,q}(\operatorname{supp} \varkappa)$

$$\iint_{\operatorname{supp} \varkappa} \psi^0(w, \gamma) f(w)\, d\sigma_w = \iint \gamma(w) \left(\frac{1}{\pi} \iint_{\operatorname{supp} \varkappa} \psi(t, w) f(t)\, d\sigma_t \right) d\sigma_w.$$

For later use we need to know of which kind the sets of zeroes of the expressions named in 4.7 are. This is stated in the following theorem (in conjunction with theorem 5.27).

5.33 Theorem. *Let $A \in L_{p,q}(\Omega, \varkappa)$ be analytic and $\not\equiv$ in Ω. If for every disk $\Delta \Subset \Omega$ there exists a closed null set N such that $\Omega \setminus N$ is simply connected and $\Delta \cap N = \emptyset$, then $L_\Omega A \neq 0$ a.e. in Ω and $L_\Omega A$ is analytic in every open subset of Ω where $\varkappa = 0$ a.e.*

Proof. Let Δ be any fixed disk $\Subset \Omega$, $\varkappa(w) = 0$ a.e. in Δ, and N be a null set satisfying the conditions mentioned in 5.33. Further, let B be a primitive function of A in $\Omega \setminus N =: \Omega_1$. B belongs to $W^1_{p,q}(\Omega_1, \varkappa)$. By theorem 5.20 every $\mathfrak{B} \in \mathfrak{K}^*_{\Omega_1}[B]$ is a solution of 5.20(I) in Ω_1. Hence $\mathfrak{B}$ and $\mathfrak{B}_w = (\mathfrak{K}^*_{\Omega_1}[B])_w$

$= L_{\Omega_1}A = L_\Omega A$ are analytic in Δ, which proves the last assertion mentioned in theorem 5.33. Now let us assume that $L_\Omega A$ is equal to zero in a subset of positive measure of Ω. Since N is a null set, $L_\Omega A = L_{\Omega_1}A$ then must be equal to zero in a subset of positive measure of Ω_1. By the representation theorem II.2.1 (cf. also theorem II.2.17) $\mathfrak{B}$ then is a constant in Ω_1, which also implies that $A \equiv 0$ in Ω. Theorem 5.33 is proved.

VI.6 FUNCTIONALS OF GRÖTZSCH–TEICHMÜLLER TYPE WITH RESPECT TO $\mathfrak{Q}(\nu, \mu)$ AND THEIR EXTREMAL FUNCTIONS

A functional χ defined on $\mathfrak{Q}(\nu, \mu)$ shall be called a functional of Grötzsch-Teichmüller type (abbreviated GT-type) with respect to $\mathfrak{Q}(\nu, \mu)$, if χ has the following properties, (PI) to (PIII).

(PI) The functional χ is upper semi-continuous on $\mathfrak{Q}(\nu, \mu)$, i.e. for any sequence f_n from $\mathfrak{Q}(\nu, \mu)$ with $f_n \to f$ in $\mathfrak{Q}(\nu, \mu)$ as $n \to \infty$ (cf. III.4.19) it holds

$$\chi[f] \geqq \limsup_{n \to \infty} \chi[f_n].$$

Hence, since $\mathfrak{Q}(\nu, \mu)$ is compact (in itself), the extremal problem

6.1 $$\chi[f] \to \max \quad \text{for } f \in \mathfrak{Q}(\nu, \mu)$$

always has at least one solution.

(PII) For every sequence of $\varkappa$-conformal boundary variations $f(z, \lambda)$ of any $f \in \mathfrak{Q}(\nu, \mu)$ with respect to $\mathfrak{c}$ and $w_0 \in \mathfrak{c} \subset \partial f(G)$ ($\mathfrak{c}$ as in VI.2)

(PII.1) $\chi[f(\cdot, \lambda)] = \chi[f] + \operatorname{Re}(\lambda \mathfrak{A}(w_0, f)) + o(\lambda)$ with $\lim\limits_{\lambda \to 0} \dfrac{o(\lambda)}{\lambda} = 0$ (if G has a non-degenerate boundary component at all). Here with every $f \in \mathfrak{Q}(\nu, \mu)$ a $\mathfrak{B}(w, f) \in W^1_{p,q}(\Omega_f, \varkappa_f)$ is associated, satisfying

(PII.2) $$\mathfrak{B}_w(w, f) = \mathfrak{A}(w, f) \quad \text{in } \Omega_f$$

and being a solution of

(PII.3) $$\mathfrak{B}_{\bar{w}}(w, f) = 2i \operatorname{Im}(\varkappa_f(w)\, \mathfrak{B}_w(w, f))$$

in Ω_f where, for every fixed $f \in \mathfrak{Q}(\nu, \mu)$, Ω_f is a domain such that

(PII.4) $$\mathbb{C} \setminus f(G) \Subset \Omega_f \quad \text{and} \quad |f(\operatorname{supp}_0(|\nu| + |\mu|) \setminus D) \setminus \Omega_f| = 0.$$

$\mathfrak{B}(w, f)$ is non-constant in Ω_f for (at least) every solution f of extremal problem 6.1; hence, for every extremal f $\mathfrak{A}(w, f) \neq 0$ a.e. in Ω_f (cf. theorem II.2.17, for instance).

(PIII) For the varied functions $f^0(z, \vartheta\gamma)$ arising by inner variation of any $f \in \mathfrak{Q}(\nu, \mu)$ with $\operatorname{supp}_0 \gamma \subseteqq f(G \setminus D)$ we have

$$\chi[f^0(\cdot, \vartheta\gamma)] = \chi[f] + \vartheta \operatorname{Re} \frac{1}{\pi} \iint \gamma(w)\, \mathfrak{A}(w, f)\, d\sigma_w + o(\vartheta), \qquad \lim_{\vartheta \to 0} \frac{o(\vartheta)}{\vartheta} = 0,$$

if $f^0(z, \vartheta\gamma)$ belongs to $\mathfrak{Q}(\nu, \mu)$ for ϑ ($\geqq 0$) sufficiently small. Here $\mathfrak{A}(w, f)$ is the function named in (PII), satisfying all the conditions mentioned there (even if $\varkappa$-conformal boundary variations are not possible in $\mathfrak{Q}(\nu, \mu)$, i.e. if G does not have any non-degenerate boundary component).

The condition for Ω_f to be a domain may be weakened. For example it is enough for Ω_f to be an open set consisting of a finite number of components and satisfying (PII.4). Then for any extremal f, $\mathfrak{B}(w,f)$ must be a non-constant solution of (PII.3) in each of these components.

The expression

6.2
$$\mathrm{d}\mathfrak{B}(w,f) \cdot \mathrm{d}w = \mathfrak{B}_w \, \mathrm{d}w^2 + \mathfrak{B}_{\bar{w}}|\mathrm{d}w|^2$$

shall be called the Grötzsch-Teichmüller quadratic differential of χ with respect to $\mathfrak{Q}(\nu, \mu)$, analogously with the conformal case. By (PII.2–4) and 1.5, $\mathfrak{B}(w,f)$ is analytic on $\partial f(G)$ (if existing). There (and in every open set $\subset \Omega$ where $\varkappa_f(w) = 0$ a.e.) we have

6.3
$$\mathrm{d}\mathfrak{B}(w,f) \cdot \mathrm{d}w = \mathfrak{A}(w,f) \, \mathrm{d}w^2,$$

and $\mathfrak{A}(w,f)$ is also analytic there (and $\not\equiv 0$ at least for extremal f).

General properties of extremal functions of extremal problems 6.1 for functionals of GT-type with respect to $\mathfrak{Q}(\nu, \mu)$ are stated in the following theorem.

6.4 Main theorem. (A) *Let χ be a functional of GT-type with respect to $\mathfrak{Q}(\nu, \mu)$, let $\mathrm{d}\mathfrak{B}(w,f) \cdot \mathrm{d}w$ be the Grötzsch-Teichmüller quadratic differential of χ, and let f_0 be any solution of the corresponding extremal problem 6.1. Then $f_0(G)$ is a slit domain, and on every boundary slit of $f_0(G)$ there holds*

(I) $\mathrm{d}\mathfrak{B}(w,f_0) \cdot \mathrm{d}w \geqq 0$.

(B) *In $G \setminus D$, $f_0(z)$ is a solution of the system*

$$\text{(II)} \quad f_{0\bar{z}} = \mu_{f_0} \cdot \overline{f_{0z}}, \qquad \mu_{f_0}(z) = (|\nu(z)| + |\mu(z)|)\frac{|\mathfrak{A}(f_0(z), f_0)|}{\mathfrak{A}(f_0(z), f_0)}.$$

This means geometrically (*cf.* section II.3):

The ratio of the main axes of the S^*-ellipses (in the image plane under the mapping f_0) corresponding to the $(0, \mu_{f_0})$-system is

$$\frac{1 + |\nu(z)| + |\mu(z)|}{1 - |\nu(z)| - |\mu(z)|} \quad \textit{a.e.}$$

The line elements $\mathrm{d}w$ *in the direction of the big semiaxes of the S^*-ellipses satisfy the relation*

$$\mathrm{d}\mathfrak{B}(w,f_0) \cdot \mathrm{d}w > 0$$

a.e. in $f_0(\operatorname{supp}_0 (|\nu(z)| + |\mu(z)|) \setminus D)$.

Here statement 6.4(I) is a short terminology for the following state of affairs:

6.4 (A′) *Every component, containing at least two points, of the complement of $f_0(G)$ (if existing) consists of the union of the closure of a finite number of analytic arcs admitting a parameter representation $w(t)$, $t \in [t_0, t_1)$, satisfying* (*cf.* 6.3) *the differential equation*

6.4 (I′) $$\mathfrak{A}(w(t), f_0)\left(\frac{\mathrm{d}w}{\mathrm{d}t}\right)^2 = 1 \quad \text{in } [t_0, t_1).$$

(Concerning the geometric shape of such curve systems, cf. [J], pp. 27–35, [S9], §§ 6 and 7.) Since $\mathfrak{A}(w, f_0)$ is analytic and $\not\equiv 0$ on $\partial f_0(G)$, i.e. it may in particular possess only a finite number of zeroes there, $\partial f_0(G)$ consists, roughly speaking, of a finite number of analytic curve arcs that may 'ramify' at a finite number of points, namely into $n + 2$ 'branches' at an n-fold zero of $\mathfrak{A}(w, f_0)$. In short, we find here essentially the same situation as with extremal problems of Grötzsch-Teichmüller type in conformal mapping. A new point of view is the occurrence of the function $\mathfrak{B}(w, f)$.)

Proof of 6.4(A). The assertion in 6.4(A) is an immediate consequence of (PII), of Schiffer's lemma in conjunction with the local behaviour of the trajectories of analytic quadratic differentials and of the variational formulae 1.22, 2.15 (whereas the variational formulae 2.16, 2.18, 2.20 serve to prove the validity of (PII) for the concrete extremal problem under consideration). Schiffer's lemma says, among other things, (see [S2]):

6.5 *Let $\mathfrak{c}$ be a bounded non-degenerate continuum whose complement is a domain, and let $A(w)$ be analytic and $\not\equiv 0$ on $\mathfrak{c}$. For each point $w_0 \in \mathfrak{c}$ let*

(1) $$\mathrm{Re}\,(\mathrm{e}^{\mathrm{i}\alpha}A(w_0)) \leqq 0$$

with every α for which there exists a sequence of conformal boundary variations $F(w, \lambda)$ with respect to $\mathfrak{c}$ and w_0 such that

(2) $$\lim_{\lambda \to 0} \frac{\lambda}{|\lambda|} = \mathrm{e}^{\mathrm{i}\alpha}.$$

Then $A(w)\,\mathrm{d}w^2 > 0$ along $\mathfrak{c}$ (i.e. $\mathfrak{c}$ is an analytic arc admitting a parameter representation $w(t)$, $t_0 \leqq t \leqq t_1$, such that $A(w(t))\,(w'(t))^2 = 1$ in $[t_0, t_1]$, or, in other words, $\mathfrak{c}$ is an analytic arc which is mapped onto a line segment parallel to the real axis by the conformal mapping $\int^{w} \sqrt{A(\zeta)}\,\mathrm{d}\zeta$).

As to the proof of 6.5 cf. also [H], [S4], [C/L].

The proof of 6.4(A) now runs as follows. Let f_0 be any extremal mapping of the extremal problem 6.1. Since $\mathfrak{B}(w, f_0)$ is non-constant in $\Omega_f \supseteq \partial f_0(G)$, $\mathfrak{A}(w, f_0)$ has only a finite number of zeroes on $\partial f_0(G)$. Let $\mathfrak{c}$ be a continuum $\subset \partial f_0(G)$ whose complement is a domain and which does not contain any zero of $\mathfrak{A}(w, f_0)$. For every sequence of Schiffer conformal boundary variations $F(w, \lambda)$ with respect to $\mathfrak{c}$ and any fixed $w_0 \in \mathfrak{c}$ for which $\frac{\lambda}{|\lambda|}$ tends to a limit $\mathrm{e}^{\mathrm{i}\alpha}$ as $\lambda \to 0$, we have, because of $f(z, \lambda) = \mathfrak{v} \circ F(f_0(z), \lambda)$ and (PII),

$$\chi[f(\cdot, \lambda)] = \chi[f_0] + \mathrm{Re}\,(|\lambda|\,\mathrm{e}^{\mathrm{i}\alpha}\mathfrak{A}(w_0, f_0)) + o(\lambda) \leqq \chi[f_0].$$

On dividing by $|\lambda|$ and then proceeding to the limit $\lambda \to 0$, we obtain 6.5(1) with $A(w_0) = \mathfrak{A}(w_0, f_0)$. Hence $\mathfrak{A}(w, f_0)\,\mathrm{d}w^2 > 0$ along $\mathfrak{c}$. Now, for every point w_0 with $\mathfrak{A}(w_0, f_0) \neq 0$, which lies on a non-degenerate continuum

$\subset \partial f_0(G)$, there exists a $\mathfrak{c}$ of the kind just considered with $w_0 \in \mathfrak{c}$. This, together with the results on the local behaviour of the trajectories of analytic quadratic differentials, gives the assertion 6.4(A).

Proof of 6.4(B) (cf. [S3], [R1], [S4]). At first we will show

6.6 $$|f_{0\bar{z}}| = |\mu_0(z)| \cdot |f_{0z}| \quad \text{a.e. in } G \setminus D, \ \mu_0 := \mu_{f_0}.$$

Let S be a set $\subset G \setminus D$ where

6.7 $$|f_{0\bar{z}}| < |\mu_0(z)| \cdot |f_{0z}|.$$

Because of the measurability of the terms on both sides of the inequality in 6.7, S is measurable in any case. Suppose S has positive measure. Let S_δ be that subset of S with

6.8 $$|f_{0\bar{z}}| < (|\mu_0(z)| - \delta) \cdot |f_{0z}| \quad \text{in } S_\delta,$$

where δ is any positive number. Of course, every S_δ is also measurable. Let δ_n be any sequence of positive numbers, $\delta_n \to 0$ as $n \to \infty$. Then $S = \bigcup_{n=1}^{\infty} S_{\delta_n}$, and because $|S| > 0$ there must be a δ_n so that $|S_{\delta_n}| > 0$. Hence there exists a positive δ so that S_δ in 6.8 has positive measure. Now we put $\Sigma := f_0(S_\delta)$. Since f_0 is a quasiconformal mapping, Σ must again have positive measure (cf. I.1.2, I.2.30). When varying f_0 by $\mathfrak{F}^0(w, \vartheta\gamma)$ with $\operatorname{supp}_0 \gamma(\cdot) \subseteqq \Sigma \subset f_0(G \setminus D)$, for $f^0(z, \vartheta\gamma) = \mathfrak{F}^0(f_0(z), \vartheta\gamma)$ we obtain

6.9
$$\left|\frac{f^0_{\bar{z}}(z, \vartheta\gamma)}{f^0_z(z, \vartheta\gamma)}\right|^2 = \left|\frac{\mathfrak{F}^0_w(w, \vartheta\gamma) f_{0\bar{z}} + \mathfrak{F}^0_{\bar{w}}(w, \vartheta\gamma)\overline{f_{0z}}}{\mathfrak{F}^0_w(w, \vartheta\gamma) f_{0z} + \mathfrak{F}^0_{\bar{w}}(w, \vartheta\gamma)\overline{f_{0\bar{z}}}}\right|^2 = \left|\frac{f_{0\bar{z}} + \vartheta \cdot \gamma(f_0(z))\overline{f_{0z}}}{f_{0z} + \vartheta \cdot \gamma(f_0(z))\overline{f_{0\bar{z}}}}\right|^2$$
$$\leqq \left(\frac{|f_{0\bar{z}}/f_{0z}| + \vartheta \cdot |\gamma(f_0(z))|}{1 - \vartheta \cdot |\gamma(f_0(z))|}\right)^2 \quad \text{a.e. in } G \setminus D,$$

and, of course, this is $\leqq |\mu_0(z)|^2 \ \forall z \in (G \setminus D) \setminus S_\delta$. For $z \in S_\delta$, by hypothesis we have $|f_{0\bar{z}}/f_{0z}| < |\mu_0(z)| - \delta$ (which incidentally means that $|\mu_0(z)| > \delta$). Hence there exists a $\vartheta_0(\delta) < 0$ such that for every $\vartheta \in (0, \vartheta_0(\delta)]$ the last expression in 6.6 also becomes $\leqq |\mu_0(z)|^2$. In particular, with

6.10 $$\gamma(w) = \frac{|\mathfrak{A}(w, f_0)|}{\mathfrak{A}(w, f_0)} \omega_\Sigma(w)$$

(where $\omega_\Sigma(\cdot)$ denotes the characteristic function of Σ and where we put $\dfrac{|\mathfrak{A}(w, f_0)|}{\mathfrak{A}(w, f_0)} = 1$ outside Ω_{f_0} and at every zero of $\mathfrak{A}(w, f_0)$) $f^0(z, \vartheta\gamma)$ belongs to $\mathfrak{Q}(\nu, \mu)$ for every $\vartheta \in (0, \vartheta_0(\delta)]$. According to (PIII) we then have

6.11 $$\chi[f^0(\cdot, \vartheta\gamma)] = \chi[f_0] + \frac{\vartheta}{\pi} \iint_\Sigma |\mathfrak{A}(w, f_0)| \, d\sigma_w + o(\vartheta).$$

Since $\mathfrak{A}(w, f_0) \neq 0$ a.e. in Ω_{f_0} and $|\Sigma| = |\Sigma \cap \Omega_{f_0}|$ (see (PII.4)) we have
$$\iint_\Sigma |\mathfrak{A}(w, f_0)| \, d\sigma_w > 0.$$

Hence

$$\chi[f^0(\cdot, \vartheta\gamma)] > \chi[f_0]$$

if $\vartheta \in (0, \vartheta_0(\delta)]$ is sufficiently small, which is in contradiction with the extremality of f_0. Assertion 6.6 is proved.

By 6.6, with f_0 being varied by $\mathfrak{F}^0(w, \vartheta\gamma)$, the corresponding varied functions $f^0(z, \vartheta\gamma)$ satisfy

6.12 $$\left|\frac{f_{\bar z}^0(z, \vartheta\gamma)}{f_z^0(z, \vartheta\gamma)}\right|^2 = \left|\frac{(f_{0\bar z}/\overline{f_{0z}}) + \vartheta\gamma}{1 + \vartheta\gamma\cdot(\overline{f_{0\bar z}/f_{0z}})}\right|^2$$

$$= |\mu_0(z)|^2 + (1 - |\mu_0(z)|^2)\cdot 2\vartheta\,\mathrm{Re}\left(\overline{\gamma(f_0(z))}\,\frac{f_{0\bar z}}{\overline{f_{0z}}}\right) + O(\vartheta^2),$$

where $|O(\vartheta^2)| \leqq M\vartheta^2 \cdot |\gamma(f_0(z))|$, M is a constant depending only on k. Now the question is under what conditions do $\gamma(w)$ exist which are not equal to zero a.e., but which on the one hand satisfy

6.13 $$\mathrm{Re}\left(\frac{f_{0\bar z}}{\overline{f_{0z}}}\,\overline{\gamma(f_0(z))}\right) \leqq \text{constant} < 0 \quad \forall z \in \mathrm{supp}_0\, \gamma(f_0(\cdot))$$

and on the other hand,

6.14 $$\mathrm{Re} \iint \gamma(w)\,\mathfrak{A}(w, f_0)\, d\sigma_w = \text{constant} > 0?$$

In view of 6.12, 6.13 aims at $\gamma(\cdot)$ yielding $f^0(z, \vartheta\gamma)$ which belong to $\mathfrak{Q}(v, \mu)$ for sufficiently small positive ϑ, and 6.14 would then imply that f_0 cannot be extremal.

To this end let $T \subset G \setminus D$ be a measurable set of positive measure where

6.15 $$f_{0\bar z} \neq |\mu_0(z)| \cdot \frac{|\mathfrak{A}(f_0(z), f_0)|}{\mathfrak{A}(f_0(z), f_0)} \cdot \overline{f_{0z}},$$

and let $T_\delta = \{z \in T : |\mu_0(z)| > \delta\}$. In the same way as with S_δ we find a $\delta > 0$ (to be fixed) such that $|T_\delta| > 0$. 6.15 means in conjunction with 6.6 that, after possible alteration of T_δ by a null set, we have

6.16 $$0 < \arg\frac{f_{0\bar z}}{\overline{f_{0z}}} - \arg\frac{|\mathfrak{A}(f_0(z), f_0)|}{\mathfrak{A}(f_0(z), f_0)} < 2\pi$$

everywhere in T_δ, under suitable choice of the branch of the argument function. We put

$$\alpha := \arg(|\mathfrak{A}(f_0(z), f_0)|/\mathfrak{A}(f_0(z), f_0)), \qquad \beta := \arg(f_{0\bar z}/\overline{f_{0z}})$$

and

6.17 $$T_{\delta\eta} := \{z \in T_\delta : 2\eta < \beta - \alpha < 2\pi - 2\eta\}.$$

Since $T_\delta = \bigcup_{n=1}^{\infty} T_{\delta\eta_n}$ if $\eta_n \to +0$ as $n \to \infty$, there exists an $\eta \in (0, \pi/2)$ such that $|T_{\delta\eta}| > 0$. When putting $T_w = f_0(T_{\delta\eta})$ and

6.18 $$\gamma(w) = e^{-i\left(\frac{\pi}{2} - \frac{\beta+\alpha}{2}\right)}\,\omega_{T_w}(w)$$

we obtain

$$\operatorname{Re}\left(\frac{f_{0\bar z}}{f_{0z}}\overline{\gamma(f_0(z))}\right) = \left|\frac{f_{0\bar z}}{f_{0z}}\right| \operatorname{Re} e^{i\left(\frac{\pi}{2}+\frac{\beta-\alpha}{2}\right)} = -\left|\frac{f_{0\bar z}}{f_{0z}}\right| \sin\frac{\beta-\alpha}{2}$$

$$\leqq -\delta \sin\eta = \text{constant} < 0$$

everywhere in $T_{\delta\eta}$. Hence, with $\gamma(w)$ defined in 6.18, $f^0(z, \vartheta\gamma)$ belongs to $\mathfrak{Q}(\nu, \mu)$ for every ϑ satisfying $0 \leqq \vartheta \leqq (1-k^2)\dfrac{2\delta}{M}\sin\eta =: \vartheta_0$. On the other hand, for these $f^0(z, \vartheta\gamma)$

$$\begin{aligned}\chi[f^0(\cdot, \vartheta\gamma)] &= \chi[f_0] + \frac{\vartheta}{\pi}\operatorname{Re}\iint\limits_{T_w} e^{-i\left(\frac{\pi}{2}-\frac{\beta+\alpha}{2}\right)} e^{-i\alpha}|\mathfrak{A}(w, f_0)|\, d\sigma_w + o(\vartheta)\\ &= \chi[f_0] + \frac{\vartheta}{\pi}\operatorname{Im}\iint\limits_{T_w} e^{i\frac{\beta-\alpha}{2}}|\mathfrak{A}(w, f_0)|\, d\sigma_w + o(\vartheta)\\ &= \chi[f_0] + \frac{\vartheta}{\pi}\iint\limits_{T_w}|\mathfrak{A}(w, f_0)|\sin\frac{\beta-\alpha}{2}\, d\sigma_w + o(\vartheta)\\ &\geqq \chi[f_0] + \frac{\vartheta\sin\eta}{\pi}\iint\limits_{T_w}|\mathfrak{A}(w, f_0)|\, d\sigma_w + o(\vartheta) > \chi[f_0],\end{aligned}$$

if ϑ is sufficiently small. This is in contradiction with the extremality of f_0. Consequently, there cannot exist a set T of positive measure as assumed in 6.15. From section II.3 we infer the geometric meaning of 6.4(II), or we can check it immediately by simple calculation. Also, 6.4(B) is proved.

6.19 *Remark.* (I) The conditions posed in 1.3 admit the special case $D = G$, $\nu = \mu = 0$ in a boundary area of G (that is, in an open set $\supset \partial G$). Then part (B) of the main theorem 6.4 loses its purpose, and the inner variations become superfluous (cf. [R2]; by the way, in the case $D = G$ it is enough to require $\mu = 0$, in place of $\nu = \mu = 0$, in a boundary area).

(II) Now let $D = G$, ν and μ are to satisfy the conditions 1.1, but no boundary area of G may exist where $\mu = 0$ (or μ = constant). The functional χ is to be defined on $\mathfrak{Q}(\nu, \mu)$ and to possess the properties (PI) and (PIII). Then we can at least say:

Each extremal function f_0 of χ maps G onto a domain whose complement has measure zero.

(Otherwise by means of inner variations $f^0(z, \vartheta\gamma)$ with $\operatorname{supp}_0 \gamma \subseteqq \mathbb{C} \setminus f_0(G)$ we would obtain a contradiction to the extremality of f_0.)

(III) The case μ = constant (also with different constants) in the boundary areas of the distinguished boundary continua can be reduced, for a large class of functionals χ (which we might again call to be of GT-type with respect to $\mathfrak{Q}(\nu, \mu)$), to the case $\mu = 0$ by means of affine transformations (for each boundary component separately).

VI.7 THE TWO EXAMPLES ONCE MORE

Now we want to apply the main theorem 6.4 to both the functionals considered in VI.4. Here we stay with the conditions posed there (which are stronger than, and, of course, contained in, those posed in 1.1). For G, which we additionally suppose to be of finite connectivity, each of its boundary components is to be non-degenerate, and $\mathrm{supp}_0\,(|\nu| + |\mu|) \setminus D$ is to have positive measure.

Both functionals are continuous on $\mathfrak{Q}(\nu, \mu)$ so that χ_1, χ_2 in any case possess the property (PI).

Under $\varkappa$-conformal boundary variation of any $f \in \mathfrak{Q}(\nu, \mu)$ χ_1 has the asymptotic expansion 4.3, that is, (PII.1) holds with

7.1 $$\mathfrak{A}(w_0, f) = 1 - \frac{1}{\pi} \iint \psi(t, w_0) \operatorname{Im}(\varkappa(t) \cdot 1)\, d\sigma_t, \quad w_0 \in \partial f(G).$$

By 5.27, with $h(w) = 1 \ \forall w \in \mathbb{C}$, which, of course, belongs to $L_{p,q}(\mathbb{C}, \varkappa)$, we have

7.2 $$\mathfrak{A}(w, f) = \mathfrak{L}_{\mathbb{C}}(1)\,(w) \quad \text{in } \mathbb{C}.$$

Hence, by 5.24 and theorem 5.20,

7.3 $\mathfrak{A}(w, f) = \mathfrak{B}_w(w, f)$, $\mathfrak{B}(w, f) \in \mathfrak{K}^*_{\mathbb{C}}[B]$ *with* $B(w) \equiv w$ *in* $\mathbb{C}$, *and* $\mathfrak{B}(w, f)$ *is equivalent to a solution of* 5.20(I) *in* $\Omega_f = \mathbb{C}$ *with* $\varkappa(\cdot) = \varkappa_f(\cdot)$; *since* $B(w) \equiv w \not\equiv$ *constant*, $\mathfrak{B}(w, f) \not\equiv$ *constant in* $\mathbb{C}$, *by theorem* 5.16.

Thus, χ_1 has the property (PII) and, by 4.4, 5.31 and 5.23, it also has the property (PIII), that is, χ_1 is of GT-type with respect to $\mathfrak{Q}(\nu, \mu)$. Before applying the main theorem we want to consider this $\mathfrak{B}(w, f)$ in more detail. Since $\varkappa = \varkappa_f$ has bounded support, $\operatorname{Im}(\varkappa B_w)$ (where $B(w) \equiv w$) belongs not only to $L_{p,q}$, but also to $L^{p,q}$, so that $\Pi^*_{\mathbb{C}}(2 \operatorname{Im}(\varkappa B_w)) = P(2 \operatorname{Im}(\varkappa B_w))_*\,(w)$ + constant (as to $(\cdot)_*$ cf. the proof of 5.18). Thus (cf. 5.7, 5.9)

7.4 $$\mathfrak{B}(w, f) = w + iP(2 \operatorname{Im} \varkappa)_*\,(w) + c = w - \frac{1}{\pi} \iint \phi(t, w) \cdot 2 \operatorname{Im} \varkappa \, d\sigma_t + c,$$

where c is an arbitraiy constant. For $c = 0$ the corresponding $\mathfrak{B}(w, f)$, then, is that uniquely determined solution of 5.20(I) in $\mathbb{C}$, which is hydrodynamically normalized and (by the representation theorem in conjunction with, for instance, the argument principle) schlicht in $\mathbb{C}$. While the boundary slits of $f_0(G)$ under an extremal mapping of the extremal problem $\operatorname{Re} a_1 \to \max$ in the conformal (respectively conformally variable) case are described by the quadratic differential,

$$d(w) \cdot dw > 0,$$

the boundary slits of $f_0(G)$ under an extremal mapping f_0 of the same extremal problem in $\mathfrak{Q}(\nu, \mu)$ are determined, according to the main theorem, by

$$d\mathfrak{B} \cdot dw > 0,$$

where $\mathfrak{B} = \mathfrak{B}(w, f_0)$ is the hydrodynamically normalized homeomorphism just mentioned of $\mathbb{C}$ onto itself. Because of the schlichtness of $\mathfrak{B}(w, f)$ (even for every $f \in \mathfrak{Q}(\nu, \mu)$) we have $\mathfrak{A}(w, f_0) = \mathfrak{B}_w(w, f_0) \neq 0$ on $\partial f_0(G)$. Thus the boundary slits are analytic arcs throughout (i.e. without branch points), as in the conformal case.

If $\mu(z)$ is real in D, so is $\varkappa_f$ in $\mathbb{C}$. Then 7.4 becomes

7.4′ $$\mathfrak{B}(w, f) \equiv w + c.$$

In this case the boundary slits are thus straight line segments parallel to the real axis, as in the conformal and quasiconformal cases (cf. [K3], theorem 2). For other special μ (with $\nu \equiv 0$) and certain special G, hyperbolic slits for instance may occur as boundary slits of the (then uniquely determined) extremal domain (cf. [R3], chap. 9, see section VI.8 below; as to further examples treated in a different way, cf. [K5]).

Let us now come to the second example, $\chi_2[f] = \operatorname{Re} f(z_0)$. In a way analogous with χ_1 we infer from the asymptotic expansion 4.3′ that the $\mathfrak{A}(w, f)$ corresponding to χ_2 is

7.5 $$\mathfrak{A}(w, f) = \mathfrak{L}_{\mathbb{C}} A(\cdot, f)(w) \quad \text{with} \quad A(w, f) = \frac{1}{f(z_0) - w} \in L_{p,q} \subset L_{p,q}(\mathbb{C}, \varkappa).$$

Let $\Omega_f = \mathbb{C} \setminus N$, where N is a closed null set containing $f(z_0)$ with $\partial f(G) \cap N = \emptyset$, and let $\mathbb{C} \setminus N$ be simply connected (for N we can take any 'polygonal arc' $\subset f(G)$ that joins $f(z_0)$ and ∞). Then $B(w, f) = -\log(w - f(z_0))$ is a (single-valued) primitive function of $A(w, f)$ in Ω_f, and every $\mathfrak{B}(w, f) \in \mathfrak{R}^*_{\Omega_f}[B(\cdot, f)]$ has the properties required in (PII) (after possible alteration on null sets). Because of 4.4′, 5.31 and 5.23, (PIII) also holds. Thus χ_2 is of GT-type with respect to $\mathfrak{Q}(\nu, \mu)$.

On account of the special assumption $\nu = \mu = 0$ for $|z| \geqq R_0$ used in our example, the $\mathfrak{B}(w, f)$ corresponding to χ_2 and $\mathfrak{Q}(\nu, \mu)$ can be characterized in a (relatively) simple manner. By the definition of $\mathfrak{R}^*_\Omega$ and because $\varkappa_f(w) = 0$ for $|w| \geqq R^*$ (cf. 1.4) we have

7.6 $$\mathfrak{B}(w, f) = -\log(w - f(z_0)) + iP\left(2 \operatorname{Im} \frac{\varkappa_f(t)}{f(z_0) - t}\right)_* (w) + c \qquad \text{a.e. in } \mathbb{C}$$

with any constant c (for instance $c = 0$), for

7.7 $\left(\frac{\varkappa_f(t)}{f(z_0) - t}\right)_*$, together with $\frac{\varkappa_f(t)}{f(z_0) - t}$, belongs to L_s $\forall s \in [1, 2)$ and has bounded support $\subseteq \{|t| \leqq R^*\}$ again; the latter we immediately infer from equation 5.8. Consequently, $\mathfrak{B}(w, f)$ from 7.6 with $c = 0$ has the following properties:

7.8 *$\mathfrak{B}(w, f) + \log(w - f(z_0))$ is (that is, can be defined as being) single-valued and continuous in $\mathbb{C} \setminus \{f(z_0)\}$.*

Indeed, $\mathfrak{B}(w, f)$ is single-valued and continous in $\mathbb{C} \setminus N$ for each $N (= N_f)$ mentioned above. Hence, the function $iP(2 \operatorname{Im} (\varkappa_f(t)/(f(z_0) - t)))_* (w)$ independent of N is single-valued and continuous on $\mathbb{C} \setminus N$ (after possible alteration on a null set) for every N. Thus, $\mathfrak{B}(w,f) + \log(w - f(z_0)$ is single-valued and continuous in $\mathbb{C} \setminus \{f(z_0)\}$.

7.9 $\mathfrak{B}(w,f) + \log(w - f(z_0)) \in L_s(K)$ *for every set* K *bounded and measurable and for every finite* $s \geqq 1$.

This follows at once from 7.7 and theorem 0.4.31. Further we have

$$\lim_{w \to \infty} \{\mathfrak{B}(w,f) + \log(w - f(z_0))\} = 0 \tag{7.10}$$

(with the branch of the logarithmus function suitably chosen), for the function $iP(2 \operatorname{Im}(\varkappa_f(t)/(f(z_0) - t)))_*(w)$ analytic for $|w| > R^*$ has a zero at ∞.

The statements 7.8 to 7.10, together with the fact that $\mathfrak{B}(w,f)$, then, is a solution of 5.20(I) in (every simply connected subdomain of) $\mathbb{C} \setminus \{f(z_0)\}$, yield:

7.11 *Let* $H(w, \zeta)$ *denote the fundamental solution of the* $(\varkappa_f, -\overline{\varkappa_f})$*-system according to* IV.3.3. *Then*

$$\mathfrak{B}(w,f) = -H(w, f(z_0)).$$

Thus, while an extremal function of the extremal problem $\operatorname{Re} f(z_0) \to \max$ in the conformal (respectively the conformally variable) case maps the domain G onto a slit domain, whose boundary slits lie on trajectories of the quadratic differential

$$d(-\log(w - f(z_0)) \cdot dw > 0,$$

an extremal function f of the same extremal problem in $\mathfrak{Q}(v, \mu)$ maps G onto a slit domain, whose boundary slits lie on trajectories of the quadratic differential

$$d(-H(w, f(z_0)) \cdot dw > 0$$

(analytic in a boundary area), where $H(w, \zeta)$ (in place of the fundamental solution $\log(w - \zeta)$ of the Cauchy–Riemann system) is now the fundamental solution of the $(\varkappa_f, -\overline{\varkappa_f})$-system (with $\varkappa_f$ depending on f, according to 1.5). As in the conformal case, these boundary slits are analytic arcs without branch points, because $H(w, f(z_0))$ admits the representation $H(w, f(z_0)) = \log(\chi(w) - \chi(f(z_0)))$, where χ is a hydrodynamically normalized quasiconformal mapping, conformal in a boundary area of $f(G)$ and in a neighbourhood of ∞, so that $H_w(w, f(z_0)) \neq 0$ especially on $\partial f(G)$.

Compared with the conformal case, extremality in $\mathfrak{Q}(v, \mu)$ seems to bear a certain additional 'degree of freedom' caused by the dependence of $\varkappa$ on f. This gives rise to the question as to whether there exists an additional necessary condition on the extremal functions, which is unknown heretofore and which disappears in the conformal case.

VI.8 A SPECIAL EXTREMAL PROBLEM

Let R be any fixed number >1, let α be any fixed number $\in [0, \pi)$ and let

$$\mu(z) = \frac{1}{R^2} e^{2i\alpha} =: \mu_0 \quad \text{when } |z| < R, \qquad \mu(z) = 0 \quad \text{otherwise.} \tag{8.1}$$

Further, let G be the domain

8.2 $$G = \overline{\mathbb{C}} \setminus \bigcup_{j=1}^{n} I_j,$$

where the I_j are closed intervals on the real axis, are mutually disjoint and are all contained in the open interval (R, R_0), where R_0 is any fixed number $>R$. For D we choose any open set with $D \supset \{|z| \leqq R\} \cup \{|z| \geqq R_0\}$. Then $\mathfrak{Q}(\mu) := \mathfrak{Q}(0, \mu)$ is constituted by all the quasiconformal mappings f of G satisfying

8.3 $$f_{\bar z} = \mu \overline{f_z} \quad \text{a.e. in } G$$

and possessing the Laurent expansion

8.4 $$f(z) = z + \frac{a_1}{z} + \frac{a_2}{z^2} + \dots$$

at the point of infinity, $a_1 = a_1(f)$. Now we consider the extremal problem

8.5 $$\chi[f] = \operatorname{Re} a_1(f) \to \max \quad \text{for } f \in \mathfrak{Q}(\mu).$$

As we have seen in the foregoing section, this χ is of GT-type with respect to $\mathfrak{Q}(\mu)$. At least one extremal function exists in $\mathfrak{Q}(\mu)$. We want to show that

8.6 $$f_0(z) = \begin{cases} z + \dfrac{e^{2i\alpha}}{z} & \text{for } |z| > R,\ z \in G, \\ z + \mu_0 \bar z & \text{for } |z| \leqq R \end{cases}$$

is the unique solution of the extremal problem 8.5.

For any $f \in \mathfrak{Q}(\mu)$ the corresponding $\varkappa = \varkappa_f$ is

8.7 $$\varkappa_f(w) = -e^{2i\alpha} \frac{R^2}{1 + R^4} \quad \text{if } w \in f(\{|z| < R\}), \qquad \varkappa_f(w) = 0 \quad \text{otherwise.}$$

When observing $f(\{|z| \leqq R\}) \subset \{|w| < 2R_0\}$ we obtain

8.8 $$\operatorname{supp} \varkappa_f \subset \{|w| < 2R_0\} \quad \text{for every } f \in \mathfrak{Q}(\mu).$$

For the corresponding $\mathfrak{B}(w,f)$ we have $\mathfrak{B}(w,f) = \mathfrak{B}_0(w,f) + c$, with

8.9 $$\mathfrak{B}_0(w,f) = w - \frac{1}{\pi} \iint \phi(t, w) \cdot 2 \operatorname{Im} \varkappa_f(t)\, d\sigma_t,$$

according to 7.4, with $\phi(t, w)$ from 5.6 and c constant. $\mathfrak{B}_0(w, f_0)$ being a solution of the $(\varkappa_{f_0}, -\overline{\varkappa_{f_0}})$-system in $\mathbb{C}$ with a simple pole at ∞ can be written down explicitely (here the geometrical interpretation from II.3 proves useful once more), namely

8.10 $$\mathfrak{B}(w) := \mathfrak{B}_0(w, f_0)$$
$$= \begin{cases} \frac{1}{2}\left[(1 + e^{-4i\alpha})\, w + (1 - e^{-4i\alpha}) \sqrt{w^2 - 4e^{2i\alpha}}\right] & \text{for } w \notin E(R, \alpha) \\ \dfrac{R^4 - e^{-4i\alpha}}{R^4 - 1}\, w + \dfrac{R^2(e^{-2i\alpha} - e^{2i\alpha})}{R^4 - 1}\, \bar w & \text{for } w \in E(R, \alpha), \end{cases}$$

where $E(R, \alpha) = f_0(\{|z| \leqq R\})$ and where we have to choose that branch of the root in 8.10 which has the expansion

$$\sqrt{w^2 - 4e^{2i\alpha}} = w + \frac{\beta}{w} + \dots \quad \text{at } \infty .$$

For the derivative $\mathfrak{B}_w(w)$ $(= \mathfrak{A}(w, f_0))$ we obtain

8.11 $$\mathfrak{B}_w(w) = \frac{1}{2}(1 + e^{-4i\alpha}) + \frac{1}{2}(1 - e^{-4i\alpha})\frac{w}{\sqrt{w^2 - 4e^{2i\alpha}}} \quad \text{for } w \notin E(R, \alpha).$$

With $w(z) = z + \dfrac{e^{2i\alpha}}{z}$ we have

8.12 $$\mathfrak{B}_w(w(z)) = \frac{z - \dfrac{e^{-2i\alpha}}{z}}{z - \dfrac{e^{2i\alpha}}{z}},$$

hence

8.13 $$|\mathfrak{B}_w(w)| = 1 \quad \text{for } w \in \partial f_0(G)$$

(note that $f_0(z)$ and $w(z)$ coincide for $z \in G$, $|z| > R$).

For $f_0(z)$ to be extremal the necessary condition

8.14 $$\mathfrak{B}_w(w(t))\, w'(t)^2 > 0 \quad \text{for } t \in I_j, \quad j = 1, 2, \dots, n,$$

must hold, where $w(t) = t + \dfrac{e^{2i\alpha}}{t}$, $t \in I_j$, is a parameter representation of the components of $\partial f_0(G)$. As we easily check by means of 8.12 this is indeed satisfied.

Now let $f_1(z)$ be any extremal function of extremal problem 8.5, where $f_1(z)$ is different from $f_0(z)$. According to the main theorem 6.4, f_1 maps G onto a slit domain whose boundary slits are unbranched analytic arcs (cf. the preceding section). Further, let

8.15 $$g(w) := f_1 \circ f_0^{-1}(w),$$

where f_0^{-1} denotes the inverse mapping of f_0. This $g(w)$ is defined and schlicht in $f_0(G)$ and has the expansion

8.16 $$g(w) = w + \frac{a_1(f_1) - a_1(f_0)}{w} + \dots$$

at ∞. Moreover, $g(w)$ is a solution of

8.17 $$g_{\bar{w}} = 2i\varkappa_{f_0} \operatorname{Im} g_w \quad \text{in } g_0(G)$$

with the $\varkappa_{f_0}$ corresponding to $f = f_0$ according to 8.7. In proving unique extremality of f_0 we have to show that

8.18 $$\operatorname{Re} b_1 < 0, \quad \text{with } b_1 = a_1(f_1) - a_1(f_0) = a_1(f_1) - e^{2i\alpha}.$$

To this end we use contour integration (cf. [K6], § 5, second proof). Let

8.19 $$h(\mathfrak{B}) := g \circ w(\mathfrak{B}),$$

$w(\mathfrak{B})$ the inverse mapping of $\mathfrak{B}(w)$. In the Laurent expansion

8.20 $$\mathfrak{B}(w) = w + \frac{\beta_1}{w} + \ldots, \quad |w| > 2R_0,$$

we have $\operatorname{Re}\beta_1 = 0$, which is a consequence of $\mathfrak{B}(w) = w + P(\mathfrak{B}_{\bar{w}})(w)$ in conjunction with $\operatorname{Re}\mathfrak{B}_{\bar{w}} = \operatorname{Re}(2\mathrm{i}\operatorname{Im}[\varkappa_{f_0}\mathfrak{B}_w]) = 0$. Thus, in the Laurent expansion

8.21 $$h(\mathfrak{B}) = \mathfrak{B} + \frac{b_1 - \beta_1}{\mathfrak{B}} + \ldots$$

we have

8.22 $$\operatorname{Re}(b_1 - \beta_1) = \operatorname{Re} b_1.$$

Besides, $h(\mathfrak{B})$ is a solution of

8.23 $$h_{\bar{\mathfrak{B}}} = 2\mathrm{i}\operatorname{Im}[\tilde{\varkappa}h_{\mathfrak{B}}] \quad \text{in } \mathfrak{B}(f_0(G))$$

with a certain $\tilde{\varkappa}(\mathfrak{B})$ (indeed $\tilde{\varkappa}(\mathfrak{B}) = \overline{\varkappa_{f_0}(w(\mathfrak{B}))}$).

By 8.22 we have,

8.24 $$\operatorname{Re} b_1 = \operatorname{Re}\left(\frac{1}{2\pi\mathrm{i}} \oint_{|\mathfrak{B}|=4R_0} h(\mathfrak{B})\,\mathrm{d}\mathfrak{B}\right) = \frac{1}{2\pi}\operatorname{Im} \oint_{|\mathfrak{B}|=4R_0} h(\mathfrak{B})\,\mathrm{d}\bar{\mathfrak{B}}.$$

By the Gauss integral theorem and 8.23, the latter is equal to

$$\frac{1}{2\pi}\operatorname{Im}\left[2\mathrm{i} \iint_{\{|\mathfrak{B}|\leq 4R_0\}\cap\mathfrak{B}(f_0(G))} h_{\bar{\mathfrak{B}}}(\mathfrak{B})\,\mathrm{d}\sigma_{\mathfrak{B}}\right] + \frac{1}{2\pi}\operatorname{Im} \oint_{\partial\mathfrak{B}(f_0(G))} h(\mathfrak{B})\,\mathrm{d}\bar{\mathfrak{B}}$$
$$= \frac{1}{2\pi}\operatorname{Im} \oint_{\partial\mathfrak{B}(f_0(G))} h(\mathfrak{B})\,\mathrm{d}\bar{\mathfrak{B}}.$$

By 8.13 and 8.14 this is

$$= \frac{1}{2\pi}\operatorname{Im} \oint_{\partial f_0(G)} g(w)\,\overline{\mathfrak{B}_w\,\mathrm{d}w} = \frac{1}{2\pi}\operatorname{Im} \oint_{\partial f_0(G)} g(w)\,|\mathfrak{B}_w|\,\mathrm{d}\bar{w} = \frac{1}{2\pi}\operatorname{Im} \oint_{\partial f_0(G)} g(w)\,\mathrm{d}\bar{w}.$$

Using

8.25

$$\operatorname{Im} \oint_{\partial f_0(G)} g\,\mathrm{d}\bar{w} = \tfrac{1}{2}\operatorname{Im} \oint_{\partial f_0(G)} \overline{(g(w) - w)}\,\mathrm{d}(g(w) - w) - \tfrac{1}{2}\operatorname{Im} \oint_{\partial f_0(G)} \overline{g(w)}\,\mathrm{d}g(w)$$

we thus obtain

8.26 $$2\pi\operatorname{Re} b_1 = \tfrac{1}{2}\operatorname{Im} \oint_{\partial f_0(G)} \overline{(g(w) - w)}\,\mathrm{d}(g(w) - w) - \tfrac{1}{2}\operatorname{Im} \oint_{\partial f_0(G)} \overline{g(w)}\,\mathrm{d}g(w).$$

By the Gauss integral theorem, respectively Green's formula, the last term on the right-hand side of 8.26 is the (non-negative) area of the complement of $f_1(G)$, and the first term is equal to

$$\tfrac{1}{2}\operatorname{Im} \oint_{|w|=r} \overline{(g(w) - w)}\,\mathrm{d}(g(w) - w) - \iint_{f_0(G)\cap\{|w|<r\}} [|(g - w)_w|^2 - |(g - w)_{\bar{w}}|^2]\,\mathrm{d}\sigma_w$$

with r sufficiently large. Because of the hydrodynamical normalization of $g(w)$ we have

8.27 $$\operatorname{Im} \oint_{|w|=r} \overline{(g - w)}\, d(g - w) = O\left(\frac{1}{r^2}\right),$$

thus

8.28 $$-\tfrac{1}{2} \operatorname{Im} \oint_{\partial f_0(G)} \overline{(g - w)}\, d(g - w) = \iint_{f_0(G)} [|(g - w)_w|^2 - |(g - w)_{\bar{w}}|^2]\, d\sigma_w.$$

Since, together with $g(w)$, $g(w) - w$ is also a solution of 8.17, the area integral in 8.28 is always

$$\geqq \iint_{f_0(G)} |g_w - 1|^2 (1 - 4|\varkappa_{f_0}(w)|^2)\, d\sigma_w.$$

Consequently,

8.29 $$2\pi \operatorname{Re} b_1 \leqq - \iint_{f_0(G)} |g_w - 1|^2 (1 - 4|\varkappa_{f_0}(w)|^2)\, d\sigma_w < 0$$

whenever $g(w) \not\equiv w$, that is, if $f_1 \not\equiv f_0$. Thus, f_0 is uniquely extremal indeed.

The boundary slits of $f_0(G)$ are slits on the hyperbolas

8.30 $$uv - (\cot 2\alpha)\, v^2 = \sin 2\alpha$$

if $\alpha \neq \pi/2$, $0 < \alpha < \pi$, and slits on the real axis in the case $\alpha = 0$ or $\alpha = \pi/2$.

The crucial point in this extremality proof is 8.13, or, what would also have been enough,

8.31 $$\operatorname{Im} \oint_{\partial f_0(G)} g(w)\, |\mathfrak{B}_w|\, d\bar{w} = \lambda \operatorname{Im} \oint_{\partial f_0(G)} g(w)\, d\bar{w}$$

with $\lambda = \lambda(f_1)$ positive (respectively non-negative, if we had to give up uniqueness). On the other hand, $\lambda(f_1) \geqq 0$ for every $f_1 \in \mathfrak{Q}(\mu)$ is not only sufficient, but also necessary for f_0 to be extremal. Thus, in certain cases this might be considered as the missing condition mentioned at the end of section VI.7. The question is, however, how to replace it by more explicit conditions, whether contained or not in the conditions for extremality stated in the main theorem.

Bibliography

[A1] L. V. Ahlfors: On quasiconformal mappings. *J. Analyse Math.*, **3** (1954), 1–58, 207–208.

[A2] L. V. Ahlfors: *Lectures on Quasiconformal Mappings.* Princeton, N. J., Toronto, New York, London: D. Van Nostrand Company, Inc., 1966.

[A/B] L. V. Ahlfors and L. Bers: Riemann's mapping theorem for variable metrics. *Ann. of Math.*, **72** (1960), 385–404.

[B1] L. Bers: An outline of the theory of pseudoanalytic functions. *Bull. Amer. Math. Soc.*, **62** (1956), 291–331.

[B2] L. Bieberbach: *Lehrbuch der Funktionentheorie, Bd. I.* Leipzig, Berlin: B. G. Teubner 1930.

[B3] L. Bieberbach: *Einführung in die konforme Abbildung.* Sammlung Göschen Bd. 768, 768a. Berlin: Walter de Gruyter & Co. 1967.

[B4] B. V. Bojarski: Generalized solutions of a system of differential equations of first order and elliptic type with discontinuous coefficients. *Mat. Sb. N. S.*, **43** (**85**) (1957), 451–503. (Russ.)

[B/N] L. Bers and L. Nirenberg: On a representation theorem for linear elliptic systems with discontinuous coefficients and its applications. *Convegno Internationale sulle Equazioni Lineari alle Derivate Parziali*, Trieste, 25–28 Agosto, 1954, pp. 111–140. Roma 1955.

[B/S] H. Behnke and F. Sommer: *Theorie der analytischen Funktionen einer komplexen Veränderlichen.* Berlin, Göttingen, Heidelberg: Springer 1955.

[C] R. Courant: *Partial Differential Equations.* (Methods of mathematical physics by R. Courant and D. Hilbert, Vol. II.) New York, London: Interscience Publishers, a division of J. Wiley & Sons, 1962.

[C/L] D. M. Campbell and J. Lamoreaux: Continua in the Plane with Limit Directions ± 1 are Line Segments. *Complex Analysis* (Proc. S.U. N.Y. Conf., Brockport, N.Y., 1976), pp. 49–52. Lecture Notes in Pure and Appl. Math., vol. 36, New York: Dekker 1978.

[F1] H. Federer: *Geometric Measure Theory.* Berlin, Heidelberg, New York: Springer 1969.

[F2] W. Franz: *Topologie.* Sammlung Göschen Bd. 1181. Berlin: Walter de Gruyter & Co. 1965.

[G1] G. M. Golusin: *Geometric Theory of Functions of a Complex Variable.* 2nd ed., Moscow: Nauka 1966. (Russ.)

[G2] H. Grötzsch: Über die Verzerrung bei nichtkonformen schlichten Abbildungen mehrfach zusammenhängender schlichter Bereiche. *Sitzungsber. Sächs. Akad. Wiss. Leipzig. Math.-Natur. Kl.*, **82** (1930), 69–80.

[G3] H. Grötzsch: Über die Verschiebung bei schlichter konformer Abbildung schlichter Bereiche. *Sitzungsber. Sächs. Akad. Wiss. Leipzig. Math.-Natur. Kl.*, **83** (1931), 254–279.

[G/L] F. W. Gehring and O. Lehto: On the total differentiability of functions of a complex variable. *Ann. Acad. Sci. Fenn. Ser. A I Math.* **272** (1959).

[G/S] P. R. Garabedian and M. Schiffer: Convexitiy of domain functionals. *J. Analyse Math.*, **2** (1953), 281–368.

[H] J. A. Hummel: *Lectures on Variational Methods in the Theory of Univalent Functions.* Lecture Note 8, University of Maryland 1972.

[J] J. A. Jenkins: *Univalent Functions and Conformal Mapping*. Ergebn. d. Math. u. ihrer Grenzgebiete, Neue Folge, Heft 18. Berlin, Göttingen, Heidelberg: Springer 1958.

[K1] S. L. Kruškal': *Quasiconformal Mappings and Riemann Surfaces*. Novosibirsk: Nauka, Sibirian division, 1975. (Russ.)

[K2] R. Kühnau: Herleitung einiger Verzerrungseigenschaften konformer und allgemeinerer Abbildungen mit Hilfe des Argumentprinzipes. *Math. Nachr.*, **39** (1969), 249–275.

[K3] R. Kühnau: Zur Methode der Randintegration bei quasikonformen Abbildungen. *Ann. Polon. Math.*, **31** (1976), 269–289.

[K4] R. Kühnau: Identitäten bei quasikonformen Normalabbildungen und eine hiermit zusammenhängende Kernfunktion. *Math. Nachr.*, **73** (1976), 73–106.

[K5] R. Kühnau: Der Flächensatz in einer Klasse schlichter Abbildungen. *Rev. Roumaine Math. Pures Appl.*, **26** (1981), 1119–1121.

[K6] R. Kühnau: Quasikonforme Abbildungen und Extremalprobleme bei Feldern in inhomogenen Medien. *J. Reine Angew. Math.*, **231** (1968), 101–113.

[L/S] M. A. Lavrent'ev and B. V. Šabat: *Methods of the Theory of Functions of a Complex Variable*. 4th ed., Moscow: Nauka 1973. (Russ.)

[L/V] O. Lehto and K. I. Virtanen: *Quasiconformal Mappings in the Plane*. 2nd ed., Berlin, Heidelberg, New York: Springer 1973.

[M1] C. B. Morrey: On the solutions of quasilinear elliptic partial differential equations. *Trans. Amer. Math. Soc.*, **43** (1938), 126–166.

[M2] C. B. Morrey: Multiple integral problems in the calculus of variations and related topics. *Ann. Scuola Norm. Pisa (III)*, **14** (1960), 1–61.

[M3] C. B. Morrey: *Multiple Integrals in the Calculus of Variations*. Berlin, Heidelberg, New York: Springer 1966.

[N] M. A. Newman: *Elements of the Topology of Plane Sets of Points*. 2nd ed. Cambridge: Cambridge University Press 1954.

[P1] A. Pfluger: Über die Äquivalenz der geometrischen und der analytischen Definition quasikonformer Abbildungen. *Comment. Math. Helv.*, **33** (1959), 23–33.

[P2] G. N. Položij: *Theory and Application of p-Analytic and (p,q)-Analytic Functions*. 2nd ed. Kiev: Naukova Dumka 1973. (Russ.)

[R1] H. Renelt: Modifizierung und Erweiterung einer Schifferschen Variationsmethode für quasikonforme Abbildungen. *Math. Nachr.*, **55** (1973), 353–379.

[R2] H. Renelt: Über Extremalprobleme für schlichte Lösungen elliptischer Differentialgleichungssysteme. *Comment. Math. Helv.*, **54** (1979), 17–41.

[R3] H. Renelt: *Über Extremalprobleme für schlichte Lösungen elliptischer Differentialgleichungssysteme*. Dissertation zur Promotion B. Martin-Luther-Universität Halle–Wittenberg 1979.

[R4] H. Renelt: Asymptotische Entwicklungen, Teichmüller-Wittichscher Verzerrungssatz und Kompaktheit in Klassen quasikonformer Abbildungen. *Ann. Acad. Sci. Fenn. Ser. A I Math.*, **6** (1981), 3–14.

[R5] H. Renelt: Über Integraltransformationen, die analytische Funktionen in Lösungen elliptischer Differentialgleichungssysteme überführen. *Ann. Polon. Math.*, **45** (1985), 1–9.

[S1] B. V. Šabat: The theorem and formula of Cauchy for quasiconformal mappings of linear classes. *Dokl. Akad. Nauk SSSR*, **49** (1949), 305–308. (Russ.)

[S2] M. Schiffer: A method of variation within the family of simple functions. *Proc. London Math. Soc.*, **44** (1938), 432–449.

[S3] M. Schiffer: A variational method for univalent quasiconformal mappings. *Duke Math. J.*, **33** (1966), 395–411.

[S4] G. Schober: *Univalent Functions—Selected Topics*. Lecture Notes in Mathematics 478. Berlin, Heidelberg, New York: Springer 1975.

[S5] W. I. Smirnow: *Lehrgang der höheren Mathematik, Teil V*. 2nd ed. Berlin: Deutscher Verlag der Wissenschaften 1967.

[S6] E. M. Stein: *Singular Integrals and Differentiability Properties of Functions*. Princeton, N.J.: Princeton University Press 1970.

[S7] S. Stoilow: *Lecons sur les principes topologiques de la théorie des fonctions analytiques*. Paris: Gauthier-Villars 1938.

[S8] K. Strebel: Ein Konvergenzsatz für Folgen quasikonformer Abbildungen. *Comment. Math. Helv.*, **44** (1969), 469–475.

[S9] K. Strebel: *Quadratic Differentials*. Ergebn. d. Math. u. ihrer Grenzgebiete, 3. Folge, Bd. 5. Berlin, Heidelberg, New York, Tokyo: Springer 1984.

[S/S1] M. Schiffer and G. Schober: A variational method for general families of quasiconformal mappings. *J. Analyse Math.*, **34** (1978), 240–264.

[S/S2] M. Schiffer and G. Schober: The dielectric Green's function and quasiconformal mapping. *J. Analyse Math.*, **36** (1979), 233–243.

[S/W] E. M. Stein and G. Weiss: *Introduction to Fourier Analysis on Euclidean Spaces*. Princeton, N.J.: Princeton University Press 1971.

[T] H. Triebel: *Interpolation Theory, Function Spaces, Differential Operators*. Berlin: Deutscher Verlag der Wissenschaften 1978, and Amsterdam, New York, Oxford: North-Holland Publ. Comp. 1978.

[V1] I. N. Vekua: The problem of reducing differential forms of elliptic type to a canonical form and the generalized Cauchy-Riemann system. *Dokl. Akad. Nauk SSSR*, **100** (1955), 197–200. (Russ.)

[V2] I. N. Vekua: *Verallgemeinerte analytische Funktionen*. Berlin: Akademie-Verlag 1963.

[V/L/A] L. I. Volkovyskij, G. L. Lunc and I. G. Aramanovič: *A Collection of Problems on Complex Analysis*. 2nd ed., Moscow: Nauka 1970. (Russ.), Engl. ed.: London: Pergamon Press 1965.

[W] W. Wendland: *Elliptic Systems in the Plane*. London, San Francisco, Melbourne: Pitman 1979.

[Y] K. Yosida: *Functional Analysis*. Berlin, Göttingen, Heidelberg: Springer 1965.

Index

absolute continuity in the sense of Tonelli 8
absolutely continuous homeomorphism 5
– equicontinuous set of homeomorphisms 5
adjoint system 62
(–)-adjoint system 62
argument principle 67
Ascoli-Arzelà theorem 29
asymptotic expansion at a finite point 70
– – at ∞ 72

Bachmann-Landau symbol 3
Beltrami coefficient 24
– solution 25
– system 24
– –, L_p-solution 24
Borel measure, regular 4
boundary variation, conformal 112
– –, $\varkappa$-conformal 114
– –, –, variational formulae 118

Cauchy integral, generalized 99
– principal value 94, 100
Cauchy-Riemann system 2
$\mathscr{C}(e_1, e_2)$ 1
chain rule I 12
– – II 12
$\mathscr{C}^k$ 2
$\mathscr{C}_0^k$ 2
$\mathscr{C}^\infty$ 2
$\mathscr{C}_0^\infty$ 2
compactness 74
– of quasiconformal mappings 32
composition rule 49
conformal boundary variation 112
convergence 74
– of quasiconformal mappings 32
$C(p)$ 16

derivative, generalized 7
– $\tau'(z)$ 5
differentiable 3
dilatation condition 24
– quotient 24

Egoroff's theorem 4
elliptic partial differential equation of second order 43
– system 41
equicontinuous 2
equidifferentiable 4
essential singularity 66

finite RBM 4
functional χ of GT-type 128, 133
– of Grötzsch-Teichmüller type 128
fundamental solution, decomposition of a 91
– – of a (ν, μ)-system 86, 89, 90, 96
– solutions, relation between 96

Gehring-Lehto theorem 6
generalized Cauchy integral 99
– – – formula 93, 96
– – – theorem 62
– derivative 7
– Poisson integral 104, 107
generalized Poisson integral formula 104
– – kernel 107
– Schwarz formula 104
Green's formula 11

Grötzsch-Teichmüller quadratic differential 129
GT-type 128

Hilbert transformation, two dimensional 15
$HL_p(D)$ 54, 78
$HL_p(D, G)$ 54
$HL_p(\{\infty\})$ 78
Hölder constant 2
– continuous 2
– exponent 2
– inequality 12
homeomorphism, absolutely continuous 5
hydrodynamically normalized 26, 110

infinitesimal ellipse 46
inner variation, variational formulae 121
integrability 3
integral operator 13
invariance of domains 68
isolated singularity 66

Jacobian 2

$\varkappa$-conformal boundary variation 114
– – – of f 118
– – –, variational formulae 118
$\mathfrak{K}_{\Omega}^{*}$ 124
$\mathfrak{K}_{\Omega}^{*-1}$ 124
K-quasiconformal mapping 24

Lebesgue's theorem 5
Liouville's theorem 68
$\mathfrak{L}_{\Omega}$ 126
$\mathfrak{L}_{\Omega}^{-1}$ 126
lower semicontinuity of the Dirichlet integral 34
L_p 2
$L_{p,\mathrm{loc}}$ 3
$L^{p,q}$ 13
$L_{p,q}$ 17
$L_{p,q}(\Omega, \varkappa)$ 125
L_p-solution 42
– of a Beltrami system 24
L_∞ 3

main theorem 110, 129
maximum principle 68
measure 2
–, outer 2
minimum principle 68
mollifier 8

norm in HL_p 54
– – L_p 3
norm in $L^{p,q}$ 13
– – $L_{p,q}$ 17
– – L_∞ 3
– – W_p^1 7
(ν, μ)-solution 42
– at a point 65
(ν, μ)-solutions, compactness 74
–, convergence 74
(ν, μ)-system 42
–, fundamental solution 86, 89, 90, 96

operator U 82, 96
outer measure 2

P 13
P_G 19
P_0 14
P_{0G} 19
P^* 18, 21
Poisson integral, generalized 104
– – formula, generalized 104
– kernel, generalized 107
pole of order n 66
(p, q)-analytic function 42
product rule 12
pseudoanalytic function of the second kind 42

$\mathfrak{Q}(\nu, \mu)$ 110
quasiconformal mappings, compactness 32, 74
– –, convergence 32, 74

RBM 4
–, finite 4
regular Borel measure 4
removable singularity 66
representation theorem 43, 46
Riemann's theorem 31
Rouché's theorem 67

saltus condition 102
Schiffer's lemma 130
schlicht mapping 23
Schwarz formula, generalized 104
S-ellipse 51
S^*-ellipse 51, 129
singularity, essential 66
–, isolated 66
–, removable 66
solution of a Beltrami system 25
– – – (ν, μ)-system 42
support of f 2
symmetry condition 103

T 15, 16
T_G 19
theorem of Gehring and Lehto 6
transformation rule 6
two-dimensional Hilbert transformation 15

uniformly continuous 2
– elliptic system 41, 43
uniqueness theorem 67

variational formulae for inner variation 121
– – – $\varkappa$-conformal boundary variation 118

Weyl's lemma 10
W_p^1 7
$W_{p,\mathrm{loc}}^1$ 8
$W_{p,q}^1(\Omega, \varkappa)$ 124
$[W_{p,q}^1(\Omega, \varkappa)]$ 124

zero of order n 65, 72